Electrical Safety Related Work Practices

Fourth Edition

Based on *NFPA 70E® 2015 Edition*

D1359798

Electrical Safety-Related Work Practices, Fourth Edition is intended to be an educational resource for the user and contains procedures commonly practiced in industry and the trade. Specific procedures vary with each task and must be performed by a qualified person. For maximum safety, always refer to specific manufacturer recommendations, insurance regulations, specific job site and plant procedures, applicable federal, state, and local regulations, and any authority having jurisdiction. The electrical training ALLIANCE assumes no responsibility or liability in connection with this material or its use by any individual or organization.

© 2015 National Joint Apprenticeship and Training Committee for the Electrical Industry

This material is for the exclusive use by the IBEW-NECA JATCs and programs approved by the NJATC. Possession and/or use by others is strictly prohibited as this proprietary material is for exclusive use by the NJATC and programs approved by the NJATC.

All rights reserved. No part of this material shall be reproduced, stored in a retrieval system, or transmitted by any means whether electronic, mechanical, photocopying, recording, or otherwise without the express written permission of the NJATC.

1 2 3 4 5 6 7 8 9 - 15 - 9 8 7 6 5 4 3 2 1

Printed in the United States of America

Acknowledgments

The NJATC wishes to thank the following companies and individuals for submitting photos for inclusion in this edition.

Boltswitch
Jim Erickson
President

Eaton's Bussmann Business
Vincent J. Saporita
Vice President Technical Sales

DuPont Engineering
Daniel Doan

H. Landis Floyd, II
Corporate Electrical Safety
Competency Leader
Principal Consultant—Electrical
Safety & Technology

Eaton Corporation
Thomas A. Domitrovich, P.E.,
LEED AP BD+C
National Application Engineer

JoAnn Frank
Marketing Communications

Fluke Corporation
Leah Friberg

IBEW Local 98
Jim Dollard, Jr.
Safety Coordinator

Ideal Industries, Inc.
Rob Conrad
Sales Training & Development
Manager

**Institute of Electrical and
Electronics Engineers (IEEE)**
Jacqueline Hansson

**International Brotherhood of
Electrical Workers (IBEW)**
Jim Spellane

Klein Tools
Stephen Ratkovich

Milwaukee Electric Tool Corp.
Scott Teson
Director of Skilled Trades

**National Electrical Contractors
Association (NECA)**
Michael J. Johnston
Executive Director Standards
and Safety

NECA
Wesley Wheeler
Director of Safety

PACE Engineers Group Pty Ltd.
Robert Fuller

Salisbury by Honeywell
John Lynch

Schneider Electric Inc.
George B. Hendricks
Manager, Learning and
Development Solutions

Service Electric Company
Mark Christian

Stark Safety Consultants
Steve Abbott

Westex by Milliken
John Gaston
Sales and Marketing Manager

The NJATC also wishes to thank the following who participated in the photos that are courtesy of NECA:

Wilson Electric Co.
Ryan Hand
Michael Maffiolli

Morse Electric, Inc.
Brad Munda
Kyle Borneman

Larry McCrae, Inc.
Jeff Costello

J.P. Rainey Company, Inc.
Dave Ganther
Bill Inforzato

Carr and Duff, Inc.
George Novelli
Tom McCusker

Northern Illinois Electrical JATC
Todd Kindred

Contents

Features

Brown **Headers** and **Subheaders** organize information within the text.

Figures, including photographs and artwork, clearly illustrate concepts from the text.

Excerpts are "ripped" from OSHA regulations or other sources.

66   Electrical Safety-Related Work Practices

Chapter 4   Introduction to Lockout, Tagging, and the Control of Hazardous Energy   67

Reenergizing Equipment

1910.333(b)(2)(v)
These requirements shall be met, in the order given, before circuits or equipment are reenergized, even temporarily.
1910.333(b)(2)(v)(A)
A qualified person shall conduct tests and visual inspections, as necessary, to verify that all tools, electrical jumpers, shorts, grounds, and other such devices have been removed, so that the circuits and equipment can be safely energized.
1910.333(b)(2)(v)(B)
Employees exposed to the hazards associated with reenergizing the circuit or equipment shall be warned to stay clear of circuits and equipment.
1910.333(b)(2)(v)(C)
Each lock and tag shall be removed by the employee who applied it or under his or her direct supervision. See Figure 4-12.

However, if this employee is absent from the workplace, then the lock or tag may be removed by a qualified person designated to perform this task provided that:
1910.333(b)(2)(v)(C)(1)
The employer ensures that the employee who applied the lock or tag is not available at the workplace, and
1910.333(b)(2)(v)(C)(2)
The employer ensures that the employee is aware that the lock or tag has been removed before he or she resumes work at that workplace.
1910.333(b)(2)(v)(D)
There shall be a visual determination that all employees are clear of the circuits and equipment.

The Control of Hazardous Energy (Lockout/Tagout)
Many lockout/tagout programs incorporate OSHA 29 CFR 1910.147. This standard is found in Subpart J, General Environmental Controls, of Part 1910, Occupational Safety and Health Standards, and is entitled "The control of hazardous energy (lockout/tagout)."
Recall that Note 2 to 1910.333(b)(2) clarifies that lockout and tagging procedures that comply with paragraphs (c) through (f) of 1910.147 will also be deemed to comply with paragraph (b)(2) of this section provided that two criteria are satisfied:
• The procedures address the electrical safety hazards covered by Subpart S.
• The procedures incorporate the requirements of paragraphs 1910.333 (b)(2)(iii)(D) and 1910.333(b)(2) (iv)(B).

OSHA Tip
Installations in electric power generation facilities that are not an integral part of, or inextricably commingled with, power generation processes or equipment are covered under § 1910.147 and Subpart S of Part 1910.

Figure 4-12. REMOVAL OF LOCKS AND TAGS

Figure 4-12. Each lock and tag is generally required to be removed by the employee who applied it. Courtesy of Ideal Industries, Inc.

1910.333(b)(2)(iii)(D)
A tag used without a lock, as permitted by paragraph (b)(2)(iii)(C) of this section, shall be supplemented by at least one additional safety measure that provides a level of safety equivalent to that obtained by use of a lock. Examples of additional safety measures include the removal of an isolating circuit element, blocking of a controlling switch, or opening of an extra disconnecting device.
1910.333(b)(2)(iv)(B)
A qualified person shall use test equipment to test the circuit elements and electrical

parts of equipment to which employees will be exposed and shall verify that the circuit elements and equipment parts are deenergized. The test shall also determine if any energized condition exists as a result of inadvertently induced voltage or unrelated voltage backfeed even though specific parts of the circuit have been deenergized and presumed to be safe. If the circuit to be tested is over 600 volts, nominal, the test equipment shall be checked for proper operation immediately after this test.

Section 1910.147 is broken down into six different major headings, each of which is indicated by a different lower-case letter:

a. Scope, application, and purpose
b. Definitions
c. General
d. Application of control
e. Release from lockout or tagout
f. Additional requirements

See 1910.147 for OSHA's requirements for "The control of hazardous energy (lockout/tagout)."

Typical Minimal Lockout Procedures
Appendix A to OSHA 29 CFR 1910.147 serves as a nonmandatory guideline to

assist employers and employees in complying with the requirements of 1910.147 and to provide other helpful information. **See Figure 4-13.** Nothing in the appendix adds to or detracts from any of the requirements of 1910.147. Rather, a simple lockout procedure is described to assist employers in developing their own procedures so they meet the requirements of this standard. For more complex systems, more comprehensive procedures may need to be developed, documented, and utilized.

Definitions
OSHA defines the terms *lockout, energy-isolating device, tagout, lockout device,* and *tagout device* in 1910.147(b) and are applicable within 1910.147:

Lockout. The placement of a lockout device on an energy-isolating device, in accordance with an established procedure, ensuring that the energy-isolating device and the equipment being controlled cannot be operated until the lockout device is removed.
Energy-isolating device. A mechanical device that physically prevents the transmission or release of energy, including but not limited to the following: A manually operated electrical circuit breaker; a disconnect switch; a manually operated switch by which the conductors of a circuit can be disconnected from all ungrounded supply conductors, and, in addition, no pole can be operated independently; a line valve; a block; and any similar device used to block or isolate energy. Push

buttons, selector switches and other control circuit type devices are not energy-isolating devices.
Tagout. The placement of a tagout device on an energy-isolating device, in accordance with an established procedure, to indicate that the energy-isolating device and the equipment being controlled may not be operated until the tagout device is removed.
Lockout device. A device that utilizes a positive means, such as a lock, either key or combination type, to hold an energy-isolating device in the safe position and prevent the energizing of a machine or equipment. Included are blank flanges and bolted slip blinds.

For additional information, visit qr.njatcdb.org Item #1561.

OSHA Tips provide supplementary information regarding OSHA standards and regulations.

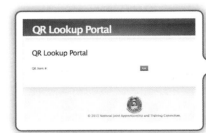

QR Lookup Portal

QR Lookup Portal

QR Item #   [Go]

© 2012 National Joint Apprenticeship and Training Committee.

For additional information, visit qr.njatcdb.org Item #1561.

Quick Response Codes (QR Codes) create a link between the textbook and the Internet. They can be scanned using Smartphone applications to obtain additional information online. (To access the information without using a Smartphone, visit qr.njatc.org and enter the referenced Item #.)

Case Studies before each chapter create a real-life context for hazardous situations.

The **Objectives** at the beginning of each chapter introduce readers to the concepts to be learned in each chapter.

At the end of each chapter, a concise chapter **Summary** and **Review Questions** reinforce the important concepts included in the text.

Case Study

A 46-year-old electrical project supervisor died when he contacted an energized conductor inside a control panel. The employer was an industrial electrical contracting company that had been in operation for 10 years. It employed 20 workers, including three electrical project supervisors. The company's written safety program, which was administered by the president/CEO and the electrical project supervisors, included disciplinary procedures. The president/CEO served as safety officer on a collateral duty basis, and the supervisors held monthly safety meetings with all crew members.

The victim had worked for the company for five years and three months as an electrical project supervisor and had approximately 27 years of electrical experience. The company and victim had been working at the packaging plant for six months before the incident; the incident was the company's first fatality.

The company had been contracted to install control cabinets, conduit, wiring, and solid-state compressor motor starters for two 400-horsepower air compressors. On the day of the incident, the victim and three coworkers (one electrical worker and two helpers) arrived at the plant at 7 a.m. They were scheduled to install the last starter and to complete the wiring from the compressor motor to the starter in the control panel, and from the starter control panel to the main distribution panel. Once installation was completed, they were to check the operation of the unit.

At approximately 3:15 p.m., the starter had been installed, and all associated wiring had been completed. The victim directed a helper to turn the switch to the "on" position at the main distribution panel, approximately six feet away, to check the starter's operation.

The helper turned the switch to the "on" position, energizing the components inside the starter control panel. The victim pushed the "start" button, and the starter indicator light activated, but the compressor motor did not start. When the compressor motor did not engage, the victim concluded that a problem existed inside the starter control panel. The victim directed the helper to retrieve a voltmeter so that he could check the continuity of the wiring inside the starter control panel. In the interim, the victim opened the starter control panel door without deenergizing the unit and reached inside to trace the wiring and check the integrity of the electrical leads. In doing so, he contacted the 480-volt primary lead for the motor starter with his left hand.

Current passed through the victim's left hand and body and exited through his feet to the ground. The victim yelled, and the helper immediately turned the main distribution switch to the "off" position as the victim collapsed to the floor. Emergency medical services (EMS) were called, and the helper checked the victim and immediately administered cardiopulmonary resuscitation (CPR). EMS personnel arrived in 10 to 15 minutes, continued CPR, and transported the victim to the local hospital, where he was pronounced dead one hour and 20 minutes after the incident occurred.

Source: For details of this case, see FACE Investigation #03CA006. Accessed June 4, 2012.

For additional information, visit qr.njatcdb.org Item #1195.

Justification, Assessment, and Implementation of Energized Work

Chapter Outline

- Application, Justification, and Documentation
- Working While Exposed to Electrical Hazards
- Approach Boundaries to Energized Electrical Conductors or Circuit Parts for Shock Protection
- Arc Flash Risk Assessment
- Other Precautions for Personnel Activities
- Personal and Other Protective Equipment
- Overhead Lines
- Underground Electrical Lines and Equipment
- Cutting or Drilling

Chapter Objectives

1. Recognize when energized work is permitted and the elements of an energized electrical work permit.
2. Understand the requirements related to approach boundaries for shock protection and those for an arc flash risk assessment.
3. Understand the requirements for personal and other protective equipment.
4. Know the requirements related to precautions for personnel activities, alerting techniques, and work within the limited approach or arc flash boundary of overhead lines, underground electrical lines and equipment, and cutting and drilling.

Review Questions

1. Incident energy analysis is defined as a component of an arc flash risk assessment used to predict the incident energy of an arc flash for a(n) __?__ set of conditions.
   a. general
   b. nominal
   c. specified
   d. unspecified

2. An energized electrical work permit is only required when work is performed within the restricted approach boundary.
   a. true
   b. false

3. Appropriate safety-related work practices must be determined before a person is exposed to the electrical hazards involved by using both a shock risk assessment and an arc flash risk assessment.
   a. true
   b. false

4. One reason why a shock risk assessment is required is to determine the personal protective equipment necessary to __?__ the possibility of electric shock.
   a. eliminate
   b. maximize
   c. minimize
   d. remove

5. Under no circumstance is an unqualified person permitted to cross the __?__ approach boundary.
   a. arc flash
   b. limited
   c. restricted
   d. secured

6. The arc flash hazard risk assessment must take into consideration the design of the overcurrent protective device and its opening time, __?__ its condition of maintenance.
   a. eliminating
   b. excluding

7. Which one of the following is not among the category of requirements related to "other precautions for personnel activities" in 130.6?
   a. Alertness, blind reaching, and confined or enclosed work spaces
   b. Conductive articles being worn, clear spaces, and doors and hinged panels
   c. Electrical hazard, arc blast, and cost benefit analysis
   d. Illumination, housekeeping duties, and anticipating failure

8. Shock protection must be worn within the __?__ approach boundary and arc flash protection must be worn within the __?__ boundary.
   a. limited, arc blast
   b. limited, arc flash
   c. restricted, limited
   d. restricted, arc flash

9. When using arc flash PPE categories to select personal protective equipment, it is important to note the parameters that must be met, such as minimum working distance, estimated short-circuit current, and __?__ fault clearing times, to determine the appropriate protective equipment.
   a. anticipated
   b. approximated
   c. maximum
   d. minimum

10. An arc flash PPE category __?__ be determined through an incident energy analysis.
   a. can
   b. cannot
   c. should
   d. should not

Summary

This chapter covered requirements related to work involving electrical hazards through overview, abbreviated content, and excerpts. Article 130 needs to be referenced to look at these requirements in their entirety.

The Article 130 requirements are broken down into ten sections covering topics including, but not limited to, those related to demonstration of justification for energized work, the energized electrical work permit, a shock hazard analysis, an arc flash hazard analysis (including equipment labeling), personal protective equipment, and other protective equipment.

These requirements, like all NFPA 70E requirements, apply within the scope of the standard spelled out in Article 90. Remember to review and apply all pertinent definitions in Article 100, the general requirements of Article 110, and the requirements for establishing an electrically safe work condition in Article 120. In addition, review and apply any relevant NFPA 70E Chapter 2 and 3 requirements and any of the related information in the Informative Annexes.

Introduction

Electrical Safety-Related Work Practices: NJATC's Guide based on NFPA 70E®

For years, many in the industry only considered electrical shock when contemplating worker protection and safe work practices. Today, however, the industry recognizes numerous additional hazards associated with work involving electrical hazards. These include, for example, the hazards associated with an arcing fault, including arc flash and arc blast. Consideration must be given to the devastating forces generated when molten copper expands to 67,000 times its original volume as it vaporizes, arc temperatures that can reach 35,000°F, pressures that can reach thousands of pounds per square foot, and shrapnel expelled from ruptured equipment at speeds that may exceed 700 miles per hour. Workers are exposed to these and other hazards even during a seemingly routine task such as voltage testing.

Working on circuits and equipment deenergized and in accordance with established lockout and tagout procedures has always been the primary safety-related work practice and a cornerstone of electrical safety. Only after it has been demonstrated that deenergizing is infeasible or would create a greater hazard, may equipment and circuit parts be worked on energized, and then only after other safety-related work practices, such as insulated tools and appropriate personal protective equipment, have been implemented. Examples of additional concerns that should be considered include worker, contractor, and customer attitude regarding energized work, comprehensiveness of an electrical safety plan, appropriate training, the role of overcurrent protective devices in electrical safety, equipment maintenance, and design and work practice considerations. These are a few of the issues that play an important role in worker safety, and are among the topics examined in this publication.

Electrical Safety-Related Work Practices has been developed in an effort to give those in the electrical industry a better understanding of a number of the hazards associated with work involving electrical hazards and the manner and conditions under which such work may be performed. These work practices and protective techniques have been developed over many years and are drawn from industry practice, national consensus standards, and federal electrical safety requirements. In many cases, these requirements are written in performance language. This publication also explores *NFPA 70E, Standard for Electrical Safety in the Workplace* as a means to comply with the electrical safety-related work practice requirements of the Occupational Safety and Health Administration.

Principal Writer

Palmer Hickman
Director of Safety, Codes and Standards
NJATC

Contributing Writer

Tim Crnko
Technical Sales
Eaton's Bussmann Business

Text Contributors

The NJATC wishes to thank the following individuals for permission to reprint materials in this edition.

Steve Abbott
Stark Safety Consultants

Vince A. Baclawski
Technical Director, Power Distribution Products
National Electrical Manufacturers Association (NEMA)

Scott Brady, PE
Western Region Manager, Technical Application Support
Eaton Corporation
Electrical Engineering Services & Systems

Electrical Construction & Maintenance (EC&M) Magazine
Penton Media

Michael Johnston
NECA
Executive Director Safety and Standards

Wesley Wheeler
NECA
Director of Safety

Electrical Safety Culture

Chapter Outline

- Safety Culture
- Hazard Awareness and Recognition
- Understanding Requirements
- Decisions

Chapter Objectives

1. Recognize the important role that a safety culture plays for every person and in every organization, and understand how it affects worker exposure to electrical hazards.
2. Understand that a number of decisions are made before and during the time a worker is exposed to electrical hazards, and appreciate how decisions can reduce or eliminate electrical hazards.
3. Recognize the important role that understanding and complying with requirements plays in reducing and eliminating hazards.

Chapter 1

References

1. National Institute for Occupational Safety and Health (NIOSH) Fatality Assessment and Control Evaluation (FACE) program
2. Occupational Safety and Health Administration (OSHA) 29 CFR Part 1910
3. OSHA 29 CFR Part 1926
4. *NFPA 70E*® 2015 edition
5. *National Electrical Code (NEC)* 2014 edition

Case Study

A 36-year-old electrician's helper was electrocuted in a fitting room of a department store located in a suburban shopping mall. The victim and a coworker were replacing the overhead fluorescent light tubes and ballast transformers in the fitting room. The employer had a written safety program that included lockout/tagout procedures and employed a safety coordinator who conducted weekly safety meetings.

Under the direction of the foreman, the victim and a second helper started work on replacing the ballasts. The foreman reportedly shut off the power to the lights by turning off and locking out the wall switch and then checked the lights with a circuit tester. This step deenergized all but one center light fixture, which was on a separate "night light" circuit that remained on. The foreman went to check the breaker but was unable to find the switch to shut off the remaining light. He confirmed that the victim had worked on live wires, and then told the victim to go ahead with the job.

The victim used a six-foot fiberglass ladder to reach the lights and began removing the tubes and ballast transformers.

He was working on the ladder when he cut the energized black wire. The power entered through the victim's hands and exited to the grounded metal doorframe that he was leaning against. The victim's coworker, who was working in the stall beside him, heard the victim say, "Help me," and saw sparks flying from the wire. The coworker cut the black wire, breaking the contact and releasing the victim, who collapsed against the metal frame.

At this time, the foreman entered the room and helped move the victim to the floor. The store manager called 911; police arrived and began cardiopulmonary resuscitation (CPR). The victim was transported to the local hospital, where he was pronounced dead.

Source: For details of this case, see New Jersey FACE Investigation #95NJ080. Accessed October 16, 2014.

For additional information, visit qr.njatcdb.org Item #1184.

INTRODUCTION

Too often a culture exists in the workplace where workers are routinely allowed and expected to work on or near energized electrical circuits. This tendency to accept the risk of an electrical injury is unacceptable and must change.

This practice might be due to ignorance of laws that have been in place for decades, lack of knowledge of the severity of the hazards, or perhaps failure to realize how quickly a task situation might change and cause an energy release. It is less likely that workers, contractors, and facilities owners would allow energized work if everyone involved in the decision-making process fully understood the laws, requirements, hazards, true costs, and consequences associated with energized work.

SAFETY CULTURE

A false sense of security devalues safe practice. As a consequence, Electrical Workers may work on energized circuits owing to misperceptions of the risks involved. These paradigms are part of the electrical work culture, and may lead workers to take risks that are not in their best interest. Many do not understand the existing and potential hazards; others, who do understand these risks, do not realize how quickly a situation can change when things go wrong. The following list of statements reflects mindsets and attitudes that can lead to taking unnecessary risks:

- I don't care what the law says—I'm going to work it energized.
- I'm an Electrical Worker; working stuff hot is part of my job.
- That's what the customer expects. If my people won't do it, then they'll get another contractor.
- It's the office of the president of the company—you can't deenergize the circuit to change that ballast.
- You can't shut that assembly line down because it will cost too much.
- There are people out of work looking for a job, so if I won't work it hot, someone else will.
- I've been doing it this way for 30 years, and nothing has ever happened to me.

- I know I should be wearing personal protective equipment (PPE), but it slows me down.
- There's no time to shut it down.
- That protective equipment is too expensive.
- What's the worst that can happen?
- It won't happen to me.

Far too many Electrical Workers believe that working on energized circuits is part of their job or is expected of them; in fact, such tasks are not part of routine electrical work. A tendency to work on or near electrical circuits while energized and accept the risk of an electrical injury creates an unacceptable culture. The need to change this mind-set must be recognized by all involved in the decision-making process. **See Figure 1-1.**

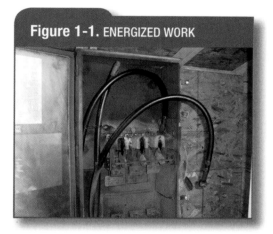

Figure 1-1. ENERGIZED WORK

Figure 1-1. *Energized work is permitted only under limited circumstances as set forth by OSHA and NFPA 70E®.* *NFPA-70E® is a registered trademark of the National Fire Protection Association, Quincy, MA.

Contractors have reported feeling pressured by their customers to work on energized equipment when a shutdown is warranted. Likewise, workers have reported feeling pressured by management to perform energized work when it is not justified. Workers who accept this risk expose themselves to injury or death. They also expose the contractors and their clients to undue risks of increased insurance premiums and loss of production. In many cases, customers may not understand the total costs and risks associated with energized work.

A well-informed client understands the hazards of energized work and the financial implications associated with an electrical incident. Equipment or circuits that are not permitted to be shut down for a few minutes ultimately might be shut down for days or weeks, or even longer, due to an unplanned event such as a dropped tool or a loose part falling into energized equipment and creating an unscheduled shutdown. A well-informed client is less likely to permit energized work, much less expect it.

HAZARD AWARENESS AND RECOGNITION

A full understanding and recognition of existing and potential hazards is crucial to ensuring that an environment is electrically safe. The following must be done at a minimum:

- Eliminate the hazard.
- Develop and implement appropriate procedures.
- Develop, conduct, and implement training for qualified and unqualified persons.
- Deenergize and follow all of the necessary steps of the lockout/tagout program and establish an electrically safe work condition unless the employer demonstrates a true need for energized work.
- Develop and implement a hazard identification and risk assessment procedure.
- Engineer out the hazards or reduce them as far as is practicable.
- Provide adequate protection against hazards when the need for energized work is demonstrated.

A comparison can be made between the hazards of driving an automobile and the hazards associated with working on or near energized electrical equipment. Protective systems such as seat belts and air bags were developed to reduce the likelihood of injury or death; likewise, personal protective equipment (PPE) was developed to increase Electrical Worker safety. Such protections have a key limitation, however; they are effective only when they are actually used.

It is much the same with the hazards associated with working while exposed to electricity. Electrical Workers will continue to be exposed to electrical hazards if they do not take appropriate steps. Potential hazards in this environment include fire, falls and falling objects, electrical shock, and the hazards associated with arcing faults, including arc flash and arc blast. An arcing fault is a fault characterized by an electrical arc through the air. Arc flash is a dangerous condition caused by the release of energy in an electric arc, usually associated with electrical distribution equipment. **See Figure 1-2.** An arcing fault, for example, could be initiated by a dropped tool or by operation of equipment that has not been maintained properly. Electrical Workers may believe that the chance of such a lapse is unlikely; however, it may need to happen only once to result in injury or death. If used, protective systems, work practices, and protective equipment can reduce or eliminate exposure to the hazards. Workers may still suffer injury or death if circuits and equipment are not worked on in an electrically safe work condition. An electrically safe work condition is defined in *NFPA 70E* Article 100 and established through the implementation of the requirements in Article 120.

The hazard of electrical shock has been recognized since the dawn of elec-

Figure 1-2. ELECTRICAL HAZARDS

Figure 1-2. Shock, arc flash, and arc blast subject workers to a number of hazards.
Courtesy of Cooper Bussmann.

Background

Historical Methods for Testing Voltage

As late as the mid-1900s, Electrical Workers performed testing for voltage procedures, employing a variety of less desirable techniques when viewed from today's perspective. On lower voltages typically found on bell, signal, and low-voltage control work, the presence of voltage (or pressure, as it commonly was called) could be tested using the "tasting method":

- A method of stripping the ends of the conductors from both sides of the circuit and placing the ends of these conductors a short distance apart on the tongue could be used to determine the presence of voltage.
- The "testee", or more appropriately, "tester" would experience a burning sensation followed by a slight salt taste. Depending on the amount of voltage present, holding one of the conductors in the bare hand and touching the other to the tongue could also be used. In this case, the body was acting as a voltage divider, lessening the burning sensation on the tongue.

Other variations of voltage testing included standing on wet ground when one end of the voltage system was grounded, while touching the tongue with the other terminal of the voltage source. This also was an "approved method." Individuals using this method often stated that once these test methods were performed, the end result was not often forgotten.

On higher voltages typically found in building power applications, the "finger method" was employed as an acceptable method of determining the presence of voltage in circuits of 250 volts or less:

- Electrical Workers would test the wires for voltage by touching the conductors to the ends of the fingers on one hand. Often, due to skin thickness, skin dryness, and calluses, the Electrical Worker would have to first lick the fingers to wet them to be able to sense the voltage being measured.

This method was billed as easy and convenient for determining whether live wires were present. The individual Electrical Worker's threshold for pain determined whether or not this was an acceptable method for everyday use. Some Electrical Workers supposedly had the ability, depending on the intensity of the sensation, to determine the actual voltage being tested.

Source: American Electrician's Handbook: A Reference Book for Practical Electrical Workers, 5th edition, by Terrell Croft (revised by Clifford C. Carr). Copyright © 1942 by The McGraw-Hill Companies, Inc. Reprinted by permission of The McGraw-Hill Companies, Inc.

tricity. The industry has evolved and made great strides to protect against electrical shock through the use of ground-fault circuit interrupters (GFCI) and rubber protective goods such as insulating gloves and blankets.

These products are effective when used and maintained properly. Even with these advances, however, injury and death still occur from electrical shock. **See Figure 1-3.** The Bureau of Labor Statistics (BLS) data for electric shocks in nonfatal cases involving days away from work for the period 1992–2001 indicate that there were an average of 2,726 cases annually in private industry.

Arc flash and arc blast constitute lesser-known hazards; electrical burns happen frequently. The BLS data for nonfatal

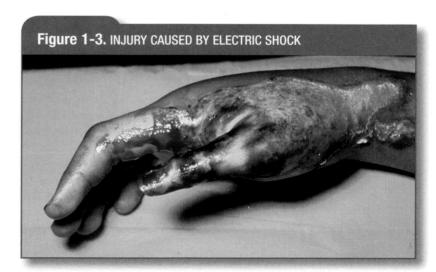

Figure 1-3. INJURY CAUSED BY ELECTRIC SHOCK

Figure 1-3. The consequences of exposure to electrical hazards are often traumatic. *Courtesy of Charles Stewart & Associates*

Background

Cardiopulmonary Resuscitation: Accepted Practice in the Early 1900s

A review of several of the techniques used as methods of resuscitation show that medical technology has come a long way since the early 1900s.

The primary method of treating an individual who had experienced heart failure and/or respiratory arrest was simple. This primary method required two rescuers to perform the resuscitation procedure:

- After placing the victim on his or her back, one rescuer would grab and wiggle the victim's tongue, while the other rescuer would work the victim's arms back and forth to help induce breathing.

While this possibly resuscitated some stricken individuals, a secondary approach was to be used should the first method fail:

- In cases where manual inflation of the lungs was attempted with no success, an attempt to cause the victim to gasp for air was performed. To initiate this gasping, the rescuers would insert two fingers into the victim's rectum, pressing them suddenly and forcibly towards the back of the individual.

Needless to say, it is not hard to understand why today's CPR methods provide more favorable results for both the victim and the rescuer.

Source: The Fire Underwriters of the United States, Standard Wiring: Electric Light and Power. H. G. Cushing Jr., New York, NY, 1911.

cases involving days away from work for the period 1992–2001 indicate an average of 1,710 electrical burns per year (peaking at 2,200 in 1995) in private industry. That averages out to nearly one worker suffering the consequences of electrical burns every hour, based on a 40-hour work week. These data were instrumental in advancing electrical safety in general and *NFPA 70E* in particular during that time period.

OSHA Tip

Occupational Safety and Health Administration (OSHA) 29 CFR 1910.333(a)(1)

Deenergized parts. Live parts to which an employee may be exposed shall be deenergized before the employee works on or near them, unless the employer can demonstrate that deenergizing introduces additional or increased hazards or is infeasible due to equipment design or operational limitations. Live parts that operate at less than 50 volts to ground need not be deenergized if there will be no increased exposure to electrical burns or to explosion due to electric arcs.

Recognizing Limitations of PPE

In addition to understanding the effects of exposure to electrical hazards, it is important to recognize that electrical protective equipment provides limited protection against electrical hazards, much like seat belts and air bags provide limited protection from the hazards that could be encountered in an automobile accident. In much the same way that a hard hat could not be expected to protect a worker from a falling steel beam, arc-rated clothing should not be expected to always allow a worker to escape an incident unscathed. An arc rating is defined in *NFPA 70E*.

Although arc-rated apparel might provide a level of protection against a thermal event that it is rated for, many other hazards might be associated with an incident. Explosive effects, including shrapnel, could rip through protective clothing, a pressure wave could rupture eardrums, or the differential pressure that results from the wave might collapse lungs and damage other internal organs. Review 130.7(A) Informational Note No. 1 and 130.7(C)(16) Informational Note

No. 2. This explanatory material advises of the limitations of their respective PPE requirements.

Understanding Requirements

It is safe to assume that not all energized work performed today falls within what the Occupational Safety and Health Administration (OSHA) recognizes as justification to work on an energized circuit. OSHA requires employers to furnish each employee with a place of employment free from recognized hazards that are causing or are likely to cause death or serious physical harm. Live parts to which an employee might be exposed must be deenergized before the employee works on or near them, unless the employer can demonstrate that deenergizing introduces additional or increased hazards or is infeasible due to equipment design or operational limitations.

It is worthwhile to consider why some laws are followed routinely while others, such as refraining from working on exposed energized electrical equipment, are often ignored. One factor that can lead to the performance of energized work is ignorance of the laws in effect. It is critical to recognize the limitations on energized work. Both *NFPA 70E* and OSHA generally stipulate that energized work is permitted only where the employer can demonstrate that deenergizing introduces additional or increased hazards or where the task to be performed is infeasible in a deenergized state due to equipment design or operational limitations. **See Figure 1-4.** Equipment must be locked out and tagged out in accordance with established policy, unless the need to work energized is demonstrated.

Electrical Workers must be intimately familiar with their company's policy on working while exposed to electrical hazards. While it may be laudable to work on energized equipment only when "we absolutely have to," that may not be entirely possible. Energized work includes voltage testing; voltage testing is among the tasks that are infeasible to perform

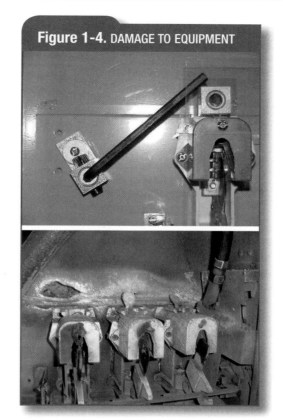

Figure 1-4. DAMAGE TO EQUIPMENT

Figure 1-4. Personnel and equipment could be severely damaged in the event of a mishap during energized work.

deenergized. As the OSHA and *NFPA 70E* requirements are fully explored in subsequent chapters, it will become apparent that the vast majority of work performed on energized equipment does not qualify as work allowed by OSHA and *NFPA 70E*.

DECISIONS

Many choices about how to perform a task are made long before a worker is placed before a hazard. Other decisions are made along the way. Many decisions are made just before, and even during, the performance of a task.

Many factors should be considered in creating a safe work environment. The following questions should be among those included in the development and implementation of a safety program,

70E Highlights

NFPA 70E includes informational notes. Per NFPA 70E Section 90.5(C), these materials are considered informational only and are not enforceable as requirements.

training, and a hazard identification and a risk assessment procedure:

- Has an electrical safety plan been developed and implemented?
- Has appropriate training been provided?
- Are safe work practices in place and understood, including lockout/tagout and placing equipment in an electrically safe work condition?
- Has the required protective equipment been provided?
- Has an attempt been made to reduce the potential worker exposure through work practice or design considerations, such as arc-resistant switchgear or remote switching?
- Was the overcurrent protection selected solely to protect the equipment or was worker protection considered as well?
- Was a current-limiting overcurrent protective device (OCPD) selected?
- Was the OCPD applied within its rating?
- If the OCPD was replaced, was the appropriate degree of current limitation applied?
- Has the OCPD been maintained properly?
- Has the impedance on the transformer on the supply side of the service changed?

These are among a few of the concerns that must be addressed when a hazard identification and risk assessment procedure is conducted.

Appropriate Priorities

A corporation's management might say that it cannot afford to develop a safety plan, provide necessary training, or provide the appropriate PPE and insulated tools. In such a case, corporate priorities should be analyzed. While there may be other important financial priorities, crucial items such as appropriate training and PPE should always be a primary consideration.

Training and Personal Responsibility

Some Electrical Workers who fail to maintain a safety culture might argue that they were never trained or were never provided with the required protective equipment. It is possible that a worker might receive training in safe work practices, but then fail to implement that training. In such a case, there is no lack of training, but the worker makes the conscious decision not to wear protective equipment. Other barriers to working in an electrically safe work condition include a lack of a safety plan and, as discussed, a lack of understanding of the hazards.

Costs of Energized Work versus Shutting Down

The true cost of an electrical injury or fatality must be considered before deciding that work must be performed on energized equipment. Consider the following:

- What is the true cost if something goes wrong?
- As mentioned earlier, a shutdown that "cannot" be scheduled could become an unscheduled shutdown. An unscheduled shutdown may ultimately cost more than a planned shutdown.
- Have the costs associated with human life, equipment, loss of production, insurance premium increases, potential exclusion from bid lists, corporate image, and worker morale been considered in the decision-making process?
- Can the equipment be shut down at night when very few, if any, people will be inconvenienced?
- Can equipment be shut down over the weekend?
- Can a shutdown be scheduled at some point?

Consider whether supervisors are rewarded for safety shortcuts. Are the costs associated with injuries and citations charged against the job, or are they considered overhead (a cost of doing business)? If such costs are not charged against the job, a job site where injuries occur may appear to be more profitable than it really is. A manager who is rewarded for safety shortcuts may not make decisions that are in the best interest of Electrical Workers or the company.

Time Pressure

Imagine an Electrical Worker operating out of a service truck who responds to a report of a transfer switch that malfunctioned at 3 a.m. at a nursing home. In such a case, the worker will be expected to get the power restored as soon as possible. An Electrical Worker must adjust the customer's understanding of "as soon as possible" to include working safely, and the added time that working safely could require. The Electrical Worker must know how to evaluate the magnitude of the hazards present, and how to make decisions about the job while simultaneously prioritizing electrical safety.

Employee Qualification

Electrical Workers must be qualified to perform the tasks to which they are assigned. An Electrical Worker must receive qualification in advance of responding to an emergency call. In such a case, items to consider include the following: Is the personal and other protective equipment in the truck? Is the protection adequate? Is the Electrical Worker qualified to make those decisions at the job site, or will someone from the engineering department need to be consulted as well? Has the worker been trained to know how to use the equipment properly and understand its limitations?

Background

INTRODUCTION TO THE NIOSH FACE PROGRAM

The National Institute for Occupational Safety and Health's (NIOSH) Fatality Assessment and Control Evaluation (FACE) program is a research program designed to identify and study all fatal occupational injuries, including those of an electrical nature. The goal of the FACE program is to prevent occupational fatalities across the United States by identifying and investigating work situations at high risk for injury and then formulating and disseminating prevention strategies to those who can intervene in the workplace.

FACE Program's Two Components

NIOSH's in-house FACE began in 1982. Participating states voluntarily notify NIOSH of traumatic occupational fatalities resulting from targeted causes of death that have included confined spaces, electrocutions, machine-related fatalities, falls from elevation, and logging incidents. The program is currently focusing on deaths associated with machinery, deaths of youths younger than 18 years of age, and street/highway construction work-zone fatalities.

The FACE program began operating as a state program in 1989. Today, nine state health or labor departments have cooperative agreements with NIOSH for conducting surveillance, targeted investigations, and prevention activities at the state level using the FACE model.

FACE is a research program; investigators do not enforce compliance with state or federal occupational safety and health standards and do not determine fault or blame.

Primary Activities of the FACE Program

The primary activities of the FACE program include the following:
- Conducting surveillance to identify occupational fatalities
- Performing investigations of specific types of events to identify injury risks
- Developing recommendations designed to control or eliminate identified risks
- Making injury prevention information available to workers, employers, and safety and health professionals

Background cont.

On-Site Investigations

On-site investigations are essential for observing sites where fatalities have occurred and for gathering facts and data from company officials, witnesses, and coworkers. Investigators collect facts and data on what was happening just before, at the time of, and right after the fatal injury. These facts become the basis for writing investigative reports.

During the on-site investigations, facts and data are collected on items such as the following:

- Type of industry involved
- Number of employees in the company
- Company safety program
- The victim's age, sex, and occupation
- The working environment
- The tasks the victim was performing
- The tools or equipment the victim was using
- The energy release that results in fatal injury
- The role of management in controlling how these factors interact

Each day, on average, 16 workers die as a result of a traumatic injury on the job. Investigations conducted through the FACE program allow the identification of factors that contribute to fatal occupational injuries. This information is used to develop comprehensive recommendations for preventing similar deaths.

FACE Information and Reports

Surveillance and investigative reports are maintained by NIOSH in a database. NIOSH researchers use this information to identify new hazards and case clusters. FACE information may suggest the need for new research or prevention efforts or for new or revised regulations to protect workers. NIOSH publications are developed to highlight high-risk work situations and to provide safety recommendations. These reports are disseminated to targeted audiences and are available on the Internet through the NIOSH homepage or through the NIOSH publications office.

The names of employers, victims, and/or witnesses are not used in written investigative reports or included in the FACE database.

Adapted from: Fatality Assessment and Control Evaluation (FACE) program website. Accessed December 1, 2014.

Summary

The list of workers' statements addressing safety culture, as presented earlier in this chapter, might seem justified and realistic. It might be true that the worker is pressured to do a job quickly or that PPE seems inconvenient or uncomfortable to wear. Nevertheless, not wearing PPE increases the potential for serious injury and perhaps—and even worse—death. When a work situation is so inherently dangerous, issues such as time pressure are irrelevant. The need to ensure the worker's safety overrides any other concerns that, while real and difficult, are not as important.

A customer's needs, while important to a business, should not be prioritized over the need of the worker to remain alive and uninjured. Remember the family and friends who will suffer the consequences of these decisions if things go wrong. Whether an incident results in an injury or a fatality, family and friends suffer emotionally and financially. If the worker does survive, he or she often requires months or years of rehabilitation.

A person might get only one chance to make a decision that those left behind will regret for years, or a lifetime. What would a loved one recommend when asked if it is worth the risk of ignoring the rules "just this one time"? There are many reasons why risks are taken. Sometimes, it could be calculated risk; at other times, it might be uninformed risk. Whatever the reason, it is not likely worth the risk.

Summary cont.

There is little question that, all too often, a culture is in place where workers are allowed and expected to work on energized equipment. Too often the existing culture supports a tendency to work routinely on or near energized electrical circuits. What is most important—that the worker performs the work without becoming injured or killed—happens only when workers, contractors, and customers become edu-cated about the hazards and ways to properly handle them. Workers, contractors, and their customers must be made aware of the requirements that are in place, the hazards that exist, the decisions that can and should be made, and the true costs associated with an incident when things go wrong. It is a multi-faceted challenge that requires a multifaceted educa-tion process and a multifaceted change in culture.

Review Questions

1. Often, a culture exists in the workplace where workers are routinely expected or allowed to work on or near energized electrical circuits. This practice might be due to ignorance of laws that have been in place for decades, lack of knowledge of the severity of the hazards, or per-haps failure to realize how __?__ a task situation might change and cause an energy release.
   a. inexplicably
   b. quickly
   c. rarely
   d. slowly

2. It is less likely that workers, contractors, and facility owners would allow energized work if __?__ involved in the decision-making process fully understood the laws and requirements, hazards and true costs, as well as consequences associated with energized work.
   a. builders
   b. employees
   c. employers
   d. everyone

3. To ensure that an environment is electrically safe, the following must be done at a mini-mum: develop, conduct, and implement __?__ for qualified and unqualified persons, develop and implement hazard identification and risk assessment procedures, and engineer out the hazards or reduce them as far as is practicable.
   a. guidelines
   b. quizzes
   c. study skills
   d. training

4. The industry has evolved and made great strides to protect personnel against electrical shock through the use of __?__ and rubber pro-tective goods such as insulating gloves and blankets.
   a. arc-fault circuit interrupters (AFCI)
   b. circuit breakers
   c. fuses
   d. ground-fault circuit interrupters (GFCI)

5. Electrical protective equipment generally does not provide protection from __?__.
   a. arc blast
   b. arc flash
   c. shock
   d. all of the above

6. An Electrical Worker must adjust the custom-er's understanding of "as soon as possible" while on a troubleshooting call to include working safely, and to the added time that working safely could require, how to evaluate the magnitude of the hazards present and how to make decisions about the job.
   a. True
   b. False

Electrical Hazard Awareness

Chapter Outline

- Workplace Hazards
- Electrical Shock
- Arcing Faults: Arc Flash and Arc Blast

Chapter Objectives

1. Identify electrical hazards.
2. Explain the effects of current on the human body.
3. Describe an arcing fault event and the effects it can have on the human body.
4. Understand the role of overcurrent protective devices in arcing fault energy release.

Chapter 2

Reference

1. *NFPA 70E*® 2015 edition

Case Study

A 19-year-old electrician's apprentice and a Journeyman Electrical Worker were installing two new switch boxes during an office building renovation project. The circuits in the room where the new switch boxes were being installed were deenergized, with the exception of the circuit to an existing metal switch box suspended by conduit from the ceiling. The circuit feeding the suspended switch box was energized by a 277-volt circuit from the adjacent room. A metal-sheathed cable fed this box and entered it through the box's side.

The Journeyman momentarily left the room and told the apprentice that "they would figure out how to wire the boxes" when he returned. Apparently, the apprentice thought the circuit feeding the suspended box was deenergized because it was in the same room as the new boxes being installed. While the apprentice was alone, he elected to disassemble the switch box suspended from the ceiling. He reached into the suspended box and cut the conductors from each of the four terminal connections in the box. Then, with his left hand, the apprentice pulled the metal sheathed conductor out of the switch box that he was holding in his right hand. The bare conductors must have contacted the box and/or his left hand. In turn, he provided a path to ground and was electrocuted. Burn marks found on the victim's right hand were consistent with the shape of the box.

The victim was found 14 feet from the switch box. Emergency medical services personnel responded and administered advanced cardiac life-support procedures. Attempts to resuscitate the victim were unsuccessful. He was pronounced dead on arrival at a nearby hospital.

Source: For details of this case, see FACE Program Case 87-34. Accessed October 16, 2014.

For additional information, visit qr.njatcdb.org Item #1186.

INTRODUCTION

Electricity is pervasive in our modern infrastructure. Electrical Workers continuously install, maintain, and troubleshoot circuits, and both they and their customers often take electricity for granted. Electricity, however, remains a very dangerous hazard for people working on or near it. Even when electrical circuits do not directly pose serious shock or burn hazards by themselves, these circuits are found adjacent to circuits with potentially lethal levels of energy. A minor shock from a low energy circuit can cause a worker to drop a tool onto another circuit, resulting in a lethal arcing fault. Involuntary reaction to a shock can result in bruises, bone fractures, and even death from collisions or falls.

The following are recognized as examples of some common electrical hazards that can cause injury and even death while a person works on or near electrical equipment and systems:

- Electrical shock
- Arc flash
- Arc blast

WORKPLACE HAZARDS

Electrical Workers face health and safety challenges on a daily basis. Potential dangers abound. As discussed in the chapter, "Electrical Safety Culture," potential reasons for the decision to expose oneself to dangers may include an incomplete understanding of the applicable safe work practices requirements. Examples may include lack of understanding of the lockout/tagout program or poor retention of hazardous communication training. Workers may lack knowledge regarding how to read a Safety Data Sheet (SDS) or be unable to recall the location and availability of the written hazardous communication program and the required list of hazardous chemicals. Potential hazards include laser equipment, paints and solvents, improperly built scaffolding, and an improperly designed excavation. Unfortunately, not all Electrical Workers exposed to these and other hazards fully appreciate the exposures they face, nor do they always know how best to avoid the many potential dangers that can cause injury and death.

Though the workplace must be evaluated to identify and eliminate all hazards, this course will focus primarily on the electrical hazards of shock, arc flash, and arc blast.

ELECTRICAL SHOCK

According to data compiled by the U.S. Department of Labor's Bureau of Labor Statistics (BLS), between the years of 2003 and 2009, electricity was the cause of death in 1,573 on-the-job fatalities. In addition, during the same period, the BLS estimates that 18,460 workers sustained nonfatal injuries in electrical incidents. Many of these nonfatal injuries took place in the manufacturing and construction industries.

Most Electrical Workers are aware of the danger of electrical shock, including electrocution. Historically, shock is the electrical hazard most prominently mentioned in the majority of electrical safety standards. In reality, few Electrical Workers truly understand that only a minimal amount of current is required to cause injury or death. The current drawn by a seven-and-a-half-watt, 120-volt lamp, passing across the chest, from hand to hand or from hand to foot, is enough to cause electrocution.

The effects of electric current on the human body depend on the following factors:

- Circuit characteristics (current, resistance, frequency, and voltage)
- Contact resistance and internal resistance of the body

OSHA Tip

Lockout is one of the main protective measures that can be taken to prevent workers from working on an energized circuit. This specific procedure involves placing a lockout device on an energy-isolating device, making it physically impossible to operate the equipment until the lockout device is removed.

OSHA defines lockout as follows:

lockout The placement of a lockout device on an energy-isolating device, in accordance with an established procedure, ensuring that the energy-isolating device and the equipment being controlled cannot be operated until the lockout device is removed.

Tagout is a protective measure in which a prominent warning device, such as a tag, is placed on the energy-isolating device, indicating that the equipment must not be operated. The tagout device warns that a worker might be injured if he or she operates the energy-isolating device.

OSHA defines tagout as follows:

tagout The placement of a tagout device on an energy-isolating device, in accordance with an established procedure, to indicate that the energy-isolating device and the equipment being controlled may not be operated until the tagout device is removed.

- The current's pathway through the body, which is determined by the contact locations and internal body chemistry
- Duration of the contact
- Environmental conditions that affect the body's contact resistance

Skin Resistance

An integral concept for understanding the magnitude of currents possible in the human body is skin contact resistance. **See Figure 2-1.** The skin's resistance can change as a function of the moisture present in its external and internal layers, which can be altered by such factors as ambient temperatures, humidity, fear, and anxiety.

Body tissue, vital organs, blood vessels, and nerve (nonfat) tissue in the human body contain water and electrolytes and are highly conductive, offering only limited resistance to alternating electric current. As the skin is broken down by electrical current, resistance drops and current levels increase.

Consider an example of a person with hand-to-hand resistance of 1,000 ohms. The voltage determines the amount of current passing through the body.

While 1,000 ohms might appear to be low, even lower levels can occur. For example, an Electrical Worker wearing sweat-soaked non-insulating gloves on both hands while maintaining a full-hand grasp of an energized conductor and a grounded pipe or conduit would approach lower levels. Moreover, cuts, abrasions, or blisters on hands can ne-

Figure 2-1. HUMAN RESISTANCE VALUES

Condition	Resistance (ohms)	
	Dry	Wet
Finger touch	40,000 to 1,000,000	4,000 to 15,000
Hand holding wire	15,000 to 50,000	3,000 to 6,000
Finger–thumb grasp	10,000 to 30,000	2,000 to 5,000
Hand holding pliers	5,000 to 10,000	1,000 to 3,000
Palm touch	3,000 to 8,000	1,000 to 2,000
Hand around 1½-inch pipe	1,000 to 3,000	500 to 1,500
Two hands around 1½-inch pipe	500 to 1,500	250 to 750
Hand immersed	N/A	200 to 500
Foot immersed	N/A	100 to 300
Human body, internal, excluding skin	200 to 1,000	

N/A: Not applicable
Data source: Kouwenhoven, W. B., and Milnor, W. R., Field Treatment of Electric Shock Cases—1, AIEE Trans. Power Apparatus and Systems, Volume 76, pp. 82–84, April 1957; discussion pp. 84–87.

Figure 2-1. *Human resistance values range for a variety of skin-contact conditions.*

gate skin resistance, leaving only internal body resistance to oppose current flow. A circuit in the range of 50 volts could be dangerous in this instance.

Using Ohm's Law, the current (I, in amperes) in a circuit can be calculated based on the circuit voltage (V) and the resistance (R). *Ohm's Law* is the mathematical relationship between voltage, current, and resistance in an electric circuit; it states that current flowing in a circuit is proportional to electromotive force (voltage) and inversely proportional to resistance: I = E/R. Current (amperes) equals voltage (volts) divided by resistance (ohms), a relationship that may be expressed in equation form:

$$I \text{ (amperes)} = \frac{V \text{ (volts)}}{R \text{ (ohms)}}$$

$$\text{Example 1: } I = \frac{480}{1,000} = 0.480 \text{ amp (480 mA)}$$

$$\text{Exmple 2: } I = \frac{120}{1,000} = 0.120 \text{ amp (120 mA)}$$

Electrical currents can cause muscles to lock up, resulting in the inability of a person to release his or her grip from the current source. The lowest current at which muscle lockup occurs is known as the let-go threshold current. The *let-go threshold* is the electrical current level at which the brain's electrical signals to muscles can no longer overcome the signals introduced by an external electrical system. Because these external signals lock muscles in the contracted position, the body may not be able to let go when the brain tells it to do so. At 60 hertz (Hz), most people have a "let-go" limit of 10 to 40 milliamperes (mA). **See Figure 2-2.**

Potential injury (current flow) also increases with time. A victim who cannot "let go" of a current source is much more likely to be electrocuted than someone whose reaction removes him or her from the circuit more quickly. A victim who is exposed for only a fraction of a second is less likely to sustain an injury.

Data addressing the levels where DC current starts flowing through the body are not presented in this publication. However, the nature of DC is that the current

Figure 2-2. ELECTRICAL SHOCK EFFECTS

Response	60 Hz, AC Current (mA)
Tingling sensation	0.5 to 3
Muscle contraction and pain	3 to 10
Let-go threshold	10 to 40
Respiratory paralysis	30 to 75
Heart fibrillation; might clamp tight	100 to 200
Tissue and organs burn	More than 1,500

Data source: Kouwenhoven, W. B., and Milnor, W. R., Field Treatment of Electric Shock Cases—1, AIEE Trans. Power Apparatus and Systems, Volume 76, pp. 82–84, April 1957; discussion pp. 84–87.

Figure 2-2. *The effects of electrical shock vary according to current level (60 Hz AC).*

remains at the same magnitude, whereas AC approaches and goes through zero 120 times a second for a 60 hertz system. Consequently, a DC system can represent a more severe shock hazard than an equivalent magnitude AC voltage system.

Extent of Injury

The most damaging paths for electrical current are through the chest cavity or head. **See Figure 2-3.** Any prolonged

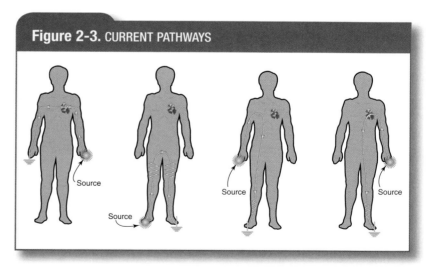

Figure 2-3. CURRENT PATHWAYS

Source

Source

Source

Source

Figure 2-3. *Current pathways through the body include hand to hand, foot to foot, right hand to right foot, and left hand to left foot.*

exposure to 60 hertz current of 10 milliamperes or more might be fatal. Fatal ventricular fibrillation of the heart (a state in which the heart does not contract, but instead twitches, stopping its rhythmic pumping action) can be initiated by a current flow of 100 to 200 milliamperes. These injuries can cause fatalities resulting from direct paralysis of the respiratory system, failure of the rhythmic heart pumping action, or immediate heart stoppage.

During fibrillation, the victim might become unconscious. Alternatively, the individual may remain conscious, deny needing help, walk a few feet, and then collapse within a short time frame. Either situation can result in death within a few minutes or hours. Prompt medical attention is needed for anyone receiving electrical shock. Many people involved in electrical incidents can be saved, provided that they receive proper medical treatment, including cardiopulmonary resuscitation (CPR), quickly.

The extent of injury resulting from electrical shock may typically not immediately be visible because the current flows through muscle tissue and organs and not through the skin, except at the entrance and exit points. Entrance and exit wounds are usually coagulated areas and may exhibit charring. Or, these areas might be missing, having "exploded" away from the body due to the level of energy present. The smaller the area of contact, the greater the heat produced. For a given current, damage in the limbs might be the greatest due to the higher current flux per unit of cross-sectional area.

Within the body, the current can burn internal body parts in its path, yet leave the skin unaffected. This type of injury might be difficult to diagnose, as the only initial signs of injury are the entry and exit wounds. Damage to the internal tissues along the current path, while not apparent immediately, might cause delayed internal tissue swelling and irritation. Prompt medical attention can minimize possible loss of blood circulation. However, for some surviving shock victims, so much internal tissue is permanently damaged that amputation of an extremity is necessary to prevent death.

Prevention

All electrocutions are preventable. A significant portion of Occupational Safety and Health Administration (OSHA) requirements is dedicated to electrical safety. Current OSHA regulations were promulgated many years ago; OSHA compliance is considered a minimum requirement for improving the safety of the workplace.

Several standards offer guidance regarding safe approach distances to minimize the possibility of shock from exposed electrical conductors of different voltage levels. One of the most recent, and perhaps the most authoritative, guidelines is presented in *NFPA 70E: Standard for Electrical Safety in the Workplace®*, in Section 130.4, Approach Boundaries to Energized Electrical Conductors or Circuit Parts for Shock Protection. (*Electrical Safety in the Workplace* and *NFPA 70E* are registered trademarks of the National Fire Protection Association, Quincy, Massachusetts.) The requirements related to approach boundaries will be covered in a subsequent chapter.

ARCING FAULTS: ARC FLASH AND ARC BLAST

The unique aspect of an arcing fault is that the fault current flows through the air between conductors or a conductor and a grounded part. The arc has an associated arc voltage because of arc impedance. The

70E Highlights

Refer to NFPA 70E Section 130.7(A) Informational Note No. 1, which provides guidance as to which hazards the PPE requirements of 130.7 are to protect against. Note even with PPE, a person could sustain burns, but they should be survivable.

product of the arcing fault current and arc voltage in a concentrated area may result in a tremendous amount of energy that is released in several forms. **See Figure 2-4**, which represents in a simple graph the basics of various types of energies released during an arcing fault. High energy release single-phase arcing faults can readily self-sustain on medium and high voltage systems. Single-phase arcing faults on systems of 600VAC or less are much more difficult to self-sustain. The lower the system voltage, the lower the probability that a single-phase arcing fault will self-sustain. The most prevalent high energy release arcing fault scenario for systems of 600 V or less is a three-phase arcing fault. However, a single-phase arcing fault on a 600VAC or less three-phase system can rapidly escalate to a three-phase arcing fault due to the resulting explosive metal vapor "cloud." DC arcing faults have a greater propensity to self-sustain than an arcing fault initiated on a single-phase AC, since DC voltage does not approach and go to zero 120 times per second as in a 60 hertz system. Therefore a DC arcing fault creates metal vapor more continuously, which increases its ability to self-sustain.

The resulting energies can take the form of intense heat, brilliant light, and tremendous pressures. Intense heat from the arcing source travels at the speed of light. The temperature of the arc terminals can reach approximately 35,000°F, which is about four times as hot as the surface of the sun. The high arc temperature changes the state of conductors from solid to both hot molten metal and metal vapor. The immediate vaporization of the conductors is an explosive change in state. Copper vapor expands to 67,000 times the volume of solid copper; thus a copper conductive component the size of a penny could expand as it vaporizes to the size of a refrigerator. Because of the expansive vaporization of conductive metal, a line-to-line or line-to-ground arcing fault can escalate into a three-phase arcing fault in less than a thousandth of a second.

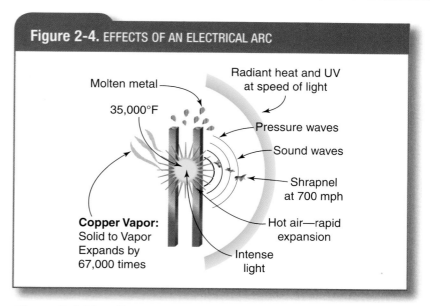

Figure 2-4. EFFECTS OF AN ELECTRICAL ARC

Molten metal

35,000°F

Radiant heat and UV at speed of light

Pressure waves

Sound waves

Shrapnel at 700 mph

Hot air—rapid expansion

Intense light

Copper Vapor: Solid to Vapor Expands by 67,000 times

Figure 2-4. The effects of an electrical arc between two copper conductors are represented by rods.

The release of intense thermal energy superheats the immediate surrounding air. The air also expands in an explosive manner. The rapid vaporization of conductors and superheating of air result in a pressure wave of air and gases and a conductive plasma cloud that can engulf a person. In addition, the thermal shock and pressures can violently destroy circuit components. The pressure waves hurl the destroyed, fragmented components like shrapnel at high velocity; shrapnel fragments can be expelled in excess of 700 miles per hour, or about the speed at which shotgun pellets leave a gun barrel. Molten metal droplets at high temperatures typically are blown out from the event due to the pressure waves.

Testing has proved that the arcing fault current magnitude and time duration are among the most critical variables in determining the amount of energy released. Notably, the predictability of arcing faults and the energy released by an arcing fault are subject to significant variance. Among the variables that affect the outcome are the following:

- Available bolted short-circuit current

Caution

It is not necessary to be in contact with the circuit to incur a serious burn. Serious or fatal burns can occur at distances of more than 10 feet from energized conductors.

- The time the fault is permitted to flow (speed of the overcurrent protective device [OCPD]; *overcurrent protective device* is a general term that includes fuses, circuit breakers, and relays.)
- Arc-gap spacing
- Size of the enclosure (or no enclosure)
- Power factor of the fault
- System voltage
- Whether the arcing fault can sustain itself
- Type of system grounding scheme
- Distance of the worker's body parts from the arc

Typically, engineering data that the industry provides about arcing faults are based on specific values of these variables. For instance, for systems of 600 volts and less, much of the data from testing on systems is with an arc-gap spacing of 1.25 or 1.0 inches and incident energy determined at 18 inches from the point of the arcing fault.

Arc Flash

Arc flash is the hazard associated with the release of thermal energy during an arcing fault. See *NFPA 70E* Article 100 for definition of arc flash hazard. In recent years, awareness of arc-flash hazards has been increasing. Each year, more than 2,000 people are admitted to burn centers in the United States with severe electrical burns. Electrical burns are considered extremely hazardous for a number of reasons. Direct contact with the circuit is not necessary to incur a serious, even deadly, burn. In fact, serious or fatal burns can occur at distances of more than 10 feet from the source of an arc flash. Ignition of flammable clothing worn by a worker is a cause of some of the most severe burns.

Molten metal splatter or thermal energy emitted from an arcing fault ignites flammable clothing, and severe burns result before the clothing can be removed.

The most common unit of measure used to quantify an arc-flash hazard is cal/cm^2. This unit of measurement is the amount of heat energy imposed on a surface area at a given distance from an arcing fault. Arc-flash suits and arc-rated clothing also have an arc rating expressed in cal/cm^2 and are designed to help protect workers from an arc-flash incident.

NFPA 70E Annex D and the current edition of IEEE 1584, *IEEE Guide for Performing Arc Flash Hazard Calculations*, by the Institute of Electrical and Electronics Engineers, illustrate methods to estimate the amount of thermal energy (incident energy) available. An arc-flash risk assessment allows a person to select the proper personal protective equipment (PPE). Incident energy analysis is discussed in detail in subsequent chapters.

Arc Blast

Arc blast is associated with the release of tremendous pressure that can occur during an arcing fault event. The worst arc blast hazards typically result from arcing faults that release high energy in a short-time duration. Various individuals and organizations in the industry have researched and continue to research ways to quantify the risks associated with arc blast. A number of the potential hazards associated with arc blast were covered in the discussion of arcing fault basics. However, there is little or no information at this time on arc-blast hazard risk assessment or on ways to protect workers from an arc-blast hazard. Arc-flash suits and other arc-rated clothing used to protect Electrical Workers from arc-flash hazards might not protect them from arc blast.

How Arcing Faults Can Affect Humans

Too few people recognize the extreme nature of electrical arcing faults, the potential for severe burns from arc flash, and the potential for injuries due to high pressures from arc blast. For Electrical Workers, the effects of an arcing fault can be devastating.

Burns are the most prevalent consequence of electrical incidents. These injuries can be due to either contact (shock hazard) or arc flash. Three basic types of such burns are distinguished:

- **Electrical burns due to current flow** – Tissue damage (whether skin deep or deeper) occurs because the body is unable to dissipate the heat from the current flow through the body. The damage to tissue can be internal and initially not obvious from external examination. Typically, electrical burns are slow to heal and frequently result in amputation.
- **Arc burns by radiant or convective heat** – Temperatures generated by electric arcs can burn flesh and ignite clothing at a distance of 10 feet or more.
- **Thermal contact burns (conductive heat)** – These injuries are normally experienced from skin contact with the hot surfaces of overheated electric conductors or a person's clothing that ignites due to an arc flash.

Studies show that when skin temperature is as low as 110°F, the body's temperature equilibrium begins to break down in about six hours. At 158°F, a one-second duration is sufficient to cause total cell destruction. Skin at temperatures of 205°F for more than 0.1 second can cause incurable, third-degree burns. **See Figure 2-5.**

In addition to burn injuries, victims of arcing faults can experience damage to their sight, hearing, and lungs, as well as skeletal, respiratory, muscular, and nervous systems. The speed of an arcing fault event can be so rapid that the human system cannot react quickly enough for a worker to take corrective measures. The radiant thermal waves, high pressure waves, spewing of hot molten metal, intense light, hurling shrapnel, and the hot conductive plasma cloud can be devastating in a fraction of a second. The intense thermal energy released can cause severe burns or ignite flammable clothing. Molten metal, when blown out from the circuit, can burn skin or ignite flammable clothing and cause serious burns over much of the body. The worker might gasp and inhale hot air and vaporized metal, sustaining severe injury to their respiratory system. The tremendous pressure blast from the vaporization of conducting materials and superheating of air can fracture ribs, collapse lungs, and knock a worker down or cause him or her to be thrown some distance.

It is important to realize that the time in which the arcing fault event runs its course might be only a fraction of a second. In a matter of approximately 0.001 second, a single-phase arcing fault can escalate to a three-phase arcing fault. Tremendous energy can be released in a few hundredths of a second. Humans cannot detect, comprehend, or react to events in these time frames.

Sometimes a greater respect for arcing fault and shock hazards is afforded to medium- and high-voltage systems. However, injury reports reveal that serious accidents are occurring at an alarming rate on systems of 600 volts or less, in part because of the high fault currents that are possible. Also, some designers, managers, and workers may not take the same necessary precautions when designing or working on low voltage systems.

Figure 2-5. SKIN TEMPERATURE AND TOLERANCE

Skin Temperature	Duration	Damage Caused
110°F	6.0 hours	Cell breakdown begins
158°F	1.0 second	Total cell destruction
176°F	0.1 second	Curable (second-degree) burn
205°F	0.1 second	Incurable (third-degree) burn

Source: Bussmann Safety BASICs Handbook, Courtesy of Cooper Bussmann, Inc. 2004.

Figure 2-5. *There is a relationship between skin temperature and tolerance.*

Figure 2-6. THRESHOLDS FOR INJURY

Threshold for Injury	Measurement
Just curable burn threshold	80°C/176°F (0.1 second)
Incurable burn threshold	96°C/205°F (just under the temperature where water will boil) for 0.1 second
Eardrum rupture threshold	720 lb/ft²
Lung damage threshold	1,728–2,160 lb/ft² (approximately the equivalent to having a compact car resting its weight on one's chest)
OSHA required ear protection threshold	85 decibel (db) for a sustained time period (Note: An increase of 3 db is equivalent to doubling the sound level.)

Source: Eaton's Bussmann Business 2014 SPD Electrical Protection Handbook.

***Figure 2-6**. There are several key thresholds for injury from an arcing fault.*

Staged Arc-Flash Tests

An ad-hoc electrical safety working group within the IEEE Petroleum and Chemical Industry Committee conducted staged arc-flash tests to investigate arcing fault hazards. These tests and others are detailed in "Staged Tests Increase Awareness of Arc-Fault Hazards in Electrical Equipment" (IEEE Petroleum and Chemical Industry Conference Record, September 1997, pp. 313–322). One finding of this IEEE paper is that current-limiting OCPDs reduce damage and arc-fault energy (provided that the fault current is within the current-limiting range of the OCPD). To better assess the benefit of limiting the current of an

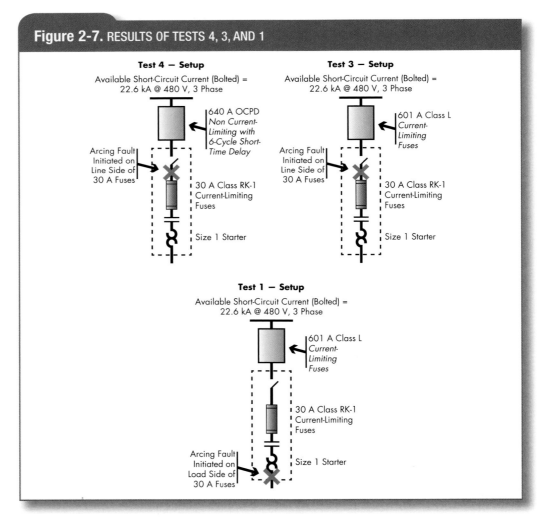

***Figure 2-7**. The one-line diagrams for Tests 4, 3, and 1 show the same available bolted short-circuit current for all three tests, but the OCPDs differ and the point of initiation of the arcing fault differs. Courtesy of Eaton's Bussmann Business.*

Figure 2-8. TEST 4

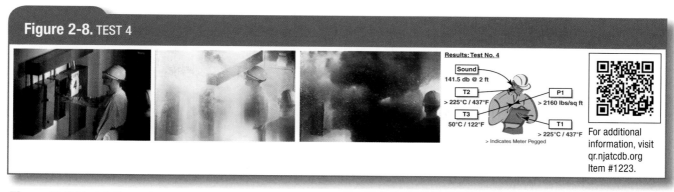

Figure 2-8. *Test 4: Staged test protected by OCPD which interrupted the fault current in six cycles (0.1 second) (not a current-limiting overcurrent protective device). Note: Unexpectedly, there was an additional fault in the wireway and the blast caused the cover to hit the mannequin in the head. Analysis results in incident energy of 5.8 cal/cm² and Arc Flash Boundary of 47 inches per 2002 IEEE 1584 (basic equations).* Courtesy of Eaton's Bussmann Business.

arcing fault, it is important to note some key thresholds of injury for humans. **See Figure 2-6.** Results of the staged arc-flash tests were recorded by sensors on mannequins and can be compared to these thresholds.

The results of three of the electrical safety working group's tests are reviewed here, identified as Test 4, Test 3, and Test 1. All three of these tests were conducted on the same electrical circuit setup with an available bolted 3-phase, short-circuit current of 22,600 symmetrical root mean square (rms) amperes at 480 volts. **See Figure 2-7.**

In each case, an arcing fault was initiated in a size 1 combination motor controller enclosure with the door open, as if an Electrical Worker were performing work on the unit while energized or be-fore it was placed in an electrically safe work condition. Test 4 and Test 3 were identical except for the OCPD protecting the circuit. In Test 4, a 640-ampere OCPD protecting the circuit cleared the test arcing fault current in six cycles. **See Figure 2-8.** In Test 3, 601-ampere (KRP-C-601SP), current-limiting fuses (Class L) are protecting the circuit; these fuses opened the test arcing fault current in less than one-half cycle and limited the current. In addition, the arcing fault was initiated on the line side of the branch circuit device in both Test 4 and Test 3 (the fault is on the feeder circuit, but within the controller enclosure). **See Figure 2-9.** In Test 1, the arcing fault is initiated on the load side of the 30-ampere branch-circuit OCPD (LPS-RK 30SP) current-limiting fuses (Class RK1).

Figure 2-9. TEST 3

Figure 2-9. *Test 3: Staged test protected by KRP-C-601SP Low-Peak™ current-limiting fuses (Class L). These fuses were in their current-limiting range and cleared in less than a ½ cycle (0.008 second). Analysis results in incident energy of 1.58 cal/cm² and arc flash boundary of 21 inches per 2002 IEEE 1584 (simplified fuse equations).* Courtesy of Eaton's Bussmann Business.

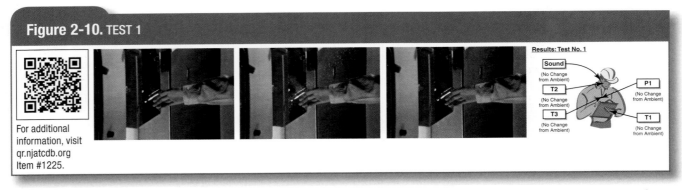

Figure 2-10. TEST 1

For additional information, visit qr.njatcdb.org Item #1225.

Figure 2-10. Test 1: Staged test protected by LPS-RK-30SP, Low-Peak™ current-limiting fuses (Class RK1). These fuses were in current-limiting range and cleared in approximately ¼ cycle (0.004 second). Analysis results in incident energy of less than 0.25 cal/cm² and Arc Flash Boundary of less than 6 inches per 2002 IEEE 1584 (simplified fuse equations). Courtesy of Eaton's Bussmann Business.

See Figure 2-10. These fuses limited this arcing fault current to a much lower amount and cleared the circuit in approximately one-fourth cycle or less.

The tests were filmed via high speed camera, and still photos were extracted. The results of the tests were recorded from the various sensors on the mannequin closest to the arcing fault. T1 and T2 recorded the temperature on the bare hand and neck, respectively. The hand with the T1 sensor was very close to the arcing fault. T3 recorded the temperature on the chest under the shirt. P1 recorded the pressure on the chest. Also, the sound level was measured at the ear. Some results "pegged the meter"—that is, the specific measurements were unable to be recorded because the actual level exceeded the range of the sensor/recorder setting. These values are shown as >, which indicates that the actual value exceeded the value given, but it is un-

known how high a level the actual value attained.

The Role of Overcurrent Protective Devices in Electrical Safety

If an arcing fault occurs while a worker is in close proximity to the fault, the survivability of the worker is mostly dependent upon the following factors:

- The magnitude of the arcing fault current
- The characteristics of the OCPDs: time to clear the arcing current and ability to limit the current
- Precautions the worker has taken prior to the event, such as wearing PPE appropriate for the hazard

The selection and performance of OCPDs play a significant role in electrical safety. Extensive tests and analysis by industry members have shown that the energy released during an arcing fault is related primarily to two characteristics of the OCPD protecting the affected circuit:

- The time it takes the OCPD to open. The faster the fault is cleared by the OCPD, the smaller the amount of energy released.
- The amount of fault current the OCPD lets through. Current-limiting OCPDs may reduce the current let-through (when the fault current

 Caution

Skin exposure to temperatures of 205°F for as little as 0.1 second can cause incurable, third-degree burns.

is within the current-limiting range of the OCPD) and, therefore, can reduce the energy released.

The lower the amount of energy released, the better for both worker safety and equipment protection. The photos and recording sensor readings from the staged tests illustrate this point.

The following conclusions can be drawn from the staged tests:

1. Arcing faults can release tremendous amounts of energy in many forms in a very short period of time, as indicated by the measured values compared to key thresholds of injury for humans. Although the circuit in Test 4 was protected by a 640-ampere OCPD, it was a non-current-limiting device and took six cycles (0.1 second) to open.

2. The OCPD's characteristic can have a significant impact on the outcome. A 601-ampere, current-limiting OCPD protected the circuit in Test 3. The current that flowed was reduced (limited), and the clearing time was one-half cycle or less. This was a significant reduction compared to Test 4. **See Figure 2-11.** Compare the Test 3 and Test 4 results to the threshold for injury values, and note the difference in exposure. In addition, note that the results of Test 1 are significantly less than those in Test 4 and even those in Test 3. The reason is that Test 1 utilized a much smaller (30-ampere) current-limiting device.

Test 3 and Test 1 both show that there are benefits of using current-limiting OCPDs. Test 1 proves the point that the greater the current limitation, the more the arcing fault energy may be reduced. Both Test 3 and Test 1 utilized very current-limiting fuses, but the lower ampere-rated fuses limited the current more than the larger ampere-rated fuses. Note that the fault current must be in the current-limiting range of the OCPD to receive the benefit of the lower current let-through. **See Figure 2-12.**

3. The shirt reduced the thermal energy exposure on the chest (the T3 sensor measured temperature under the shirt). This illustrates the benefit of workers wearing protective garments.

Figure 2-11. COMPARISON OF TESTS

Test	Protective Device Used	OCPD Clearing Time
Test 4	640-ampere, non-current-limiting device	Six cycles
Test 3	KRP-C 601SP, 601-ampere, current-limiting fuses (Class L)	Less than one-half cycle
Test 1	LPS-RK 30SP, 30-ampere, current-limiting fuses (Class RK1)	One-fourth cycle

Figure 2-11. Three staged arc-flash tests are compared.

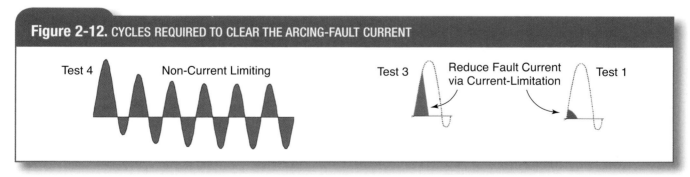

Figure 2-12. CYCLES REQUIRED TO CLEAR THE ARCING-FAULT CURRENT

Test 4 — Non-Current Limiting

Test 3 — Reduce Fault Current via Current-Limitation — Test 1

Figure 2-12. The oscillographs of Test 4, Test 3, and Test 1 show the number of cycles required to clear the arcing-fault current.

Summary

Recognizing the many hazards to which a worker might be exposed and understanding the severe consequences of that exposure are important steps in convincing everyone involved that much must be done to provide a workplace free from recognized hazards. In summary, the electrical hazards associated with working on or near exposed energized electrical conductors include the following:

- Electrical shock
- Arc flash
- Arc blast

Electrical shock hazard is related to the magnitude of current and the path through the human body. Arc-flash hazard is related to the arcing-current magnitude and time duration. Overcurrent protection can have a significant role in the level of arc-flash hazard. At this time, there is insufficient information to quantify arc-blast hazards and recommend appropriate PPE.

Review Questions

1. Which of the following electrical hazards can potentially cause injury and even death to a person working on or near electrical equipment and systems?
   a. Arc blast
   b. Arc flash
   c. Electric shock
   d. All of the above

2. Few Electrical Workers understand that a minimal amount of current is required to cause injury or death due to shock hazard. What is the approximate range of current needed to cause the onset of heart fibrillation?
   a. 5 to 10 mA
   b. 10 to 20 mA
   c. 30 to 50 mA
   d. 100 to 200 mA

3. What does the severity of injury due to electric current flowing through the human body depend on?
   a. The amount of current through the body
   b. The current's pathway through the body
   c. Time duration of the contact
   d. All of the above

4. The most damaging path for electrical current is through the chest cavity and head. Ventricular fibrillation of the heart can cause fatalities resulting from __?__.
   a. contractions of the digestive system
   b. failure of the rhythmic heart pumping action
   c. indirect contact
   d. tingling sensation

5. All electrocutions are preventable. Which standard offers guidance regarding safe approach distances for different voltage levels to minimize the possibility of shock from exposed electrical conductors?
   a. *NFPA 70*
   b. *NFPA 70E*
   c. *NFPA 72*
   d. *NFPA 99*

6. The unique aspect of an arcing fault is that the resulting energies can be in the form of intense heat, brilliant light, and tremendous pressures. Intense heat from the arcing source travels at the speed of light. The temperature of the arc terminals can reach approximately __?__, or about four times as hot as the surface of the sun.
   a. 650°F
   b. 800°F
   c. 1,900°F
   d. 35,000°F

7. Incident energy is the amount of energy impressed on a surface, a certain distance from the source, generated during an electrical arc event. One of the units used to measure incident energy is __?__.
   a. amperes per square yard
   b. calories per centimeter squared (cal/cm^2)
   c. pounds per square foot
   d. watts per cubic foot

8. Bare skin exposed to an arc-flash event can result in injury due to thermal conditions. Skin exposed to temperatures of __?__ for more than one-tenth of one second can cause incurable, third-degree burns.
   a. 110°F
   b. 158°F
   c. 176°F
   d. 205°F

9. If an arcing fault occurs while a worker is in close proximity to the fault, the survivability of the worker is mostly dependent upon which of the following factors?
   a. Characteristics of the overcurrent protective device
   b. Magnitude of the arcing current
   c. Precautions the worker has taken prior to the event
   d. All of the above

OSHA Considerations

Chapter Outline

- OSHA: History, Application, Enforcement, and Responsibility
- The OSH Act and OSHA Standards and Regulations
- Safety and Health Regulations for Construction
- Occupational Safety and Health Standards

Chapter Objectives

1. Demonstrate an understanding of OSHA's history, application, enforcement, and responsibility.
2. Demonstrate an understanding of the OSH Act, the General Duty Clause, and OSHA Standards and Regulations.
3. Understand the regulations presented in Part 1926, Safety and Health Regulations for Construction.
4. Understand the regulations presented in Part 1910, Occupational Safety and Health Standards.

Chapter 3

References

1. *NFPA 70E®*, 2015 Edition
2. OSHA 29 CFR Part 1926
3. OSHA 29 CFR Part 1910
4. Occupational Safety and Health Act of 1970

Case Study

A crew of electricians was working at a facility that was shut down for the July 4 holiday. The crew members had approximately one hour of work left to finish before they could go home and enjoy the holiday. The workers were pulling three sets of wiring from a source in the main plant to new electrical equipment in an addition to the facility. Two sets were for air conditioning, and one set was for a new lighting panel. Each set of wiring had its own breaker, which the foreman—but not the lead electrician—had locked out, modifying the normally followed lockout/tagout procedure. Normally, the employee performing the work would place his or her lockout/tagout equipment on the breakers and then remove the lockout/tagout equipment after the work was completed.

After completing connections for the new lighting panel, the lead electrician was getting ready to connect the wires for the air conditioning. As he pulled the wires into a junction box, he tapped the ends of the wires into his right hand to make them even. He was not wearing insulated gloves as he handled the wires and made the connections. At the same time, the lead electrician was ready for the breaker to the lighting panel to be turned on and instructed the foreman to throw the breaker to the "on" position. The foreman, thinking he should throw all three breakers to the "on" position, walked over to the breaker panel and removed his lockout/tagout on all three breakers. He then proceeded to throw all three to the "on" position. This action sent electricity through the wires into the lead electrician's hand, killing him.

Reportedly, the victim looked at his coworker, said "Help me," and then collapsed. Nearby workers called out to the foreman to contact emergency services, which he did immediately. While emergency services were en route, cardiopulmonary resuscitation was performed until paramedics arrived. Paramedics took the victim to a nearby hospital, where a physician notified the coroner, who declared the victim dead. The cause of death was electrocution.

Source: For details of this case, see FACE Investigation 03KY115. Accessed October 16, 2014.

For additional information, visit qr.njatcdb.org Item #1187.

INTRODUCTION

The Occupational Safety and Health Administration (OSHA) was established more than four decades ago when the U.S. Congress passed the Occupational Safety and Health Act (OSH Act) of 1970. Its purpose, in part, is "to assure for far as possible every working man and woman in the nation safe and healthful working conditions and preserve our human resources." Requirements were put into place that, when complied with, go a long way toward avoiding the hazards encountered in the workplace.

Electrical Workers are exposed to a number of hazards, and the work that they perform may be covered by any number of OSHA regulations. The primary focus of this chapter is on the regulations in Part 1926, Safety and Health Regulations for Construction, and Part 1910, Occupational Safety and Health Standards. The discussion here provides and overview of a number of performance-based OSHA regulations that may apply.

The information discussed in this chapter is not intended to be all-inclusive or the basis for an electrical safety program. Instead, the intent is to point out a number of the regulations that may apply. Part 1926 and Part 1910 regulations must be studied in their entirety for a full and complete look at these provisions and their application.

OSHA: HISTORY, APPLICATION, ENFORCEMENT, AND RESPONSIBILITY

OSHA publication 3302-09R 2014, entitled *All About OSHA*, includes the following general overview of basic topics related to OSHA's mission, what it covers, and how it operates.

On December 29, 1970, President Nixon signed the Occupational Safety and Health Act of 1970 (OSH Act) into law, establishing OSHA. Congress created OSHA to assure safe and healthful conditions for working men and women by setting and enforcing standards and providing training, outreach, education, and compliance assistance. Coupled with the efforts of employers, workers, safety and health professionals, unions, and advocates, OSHA and its state partners have dramatically improved workplace safety, reducing work-related deaths and injuries by more than 65%.

History

According to OSHA, in 1970, an estimated 14,000 workers were killed on the job—about 38 every day. For 2010, the Bureau of Labor Statistics reports this number fell to about 4,500, or about 12 workers per day. At the same time, U.S. employment has almost doubled to over 130 million workers at more than 7.2 million worksites. The rate of reported serious workplace injuries and illnesses has also dropped markedly, from 11 per 100 workers in 1972 to 3.5 per 100 workers in 2010.

OSHA's safety and health standards have prevented countless work-related injuries, illnesses, and deaths. Nevertheless, far too many preventable injuries and fatalities continue to occur. Significant hazards and unsafe conditions still exist in U.S. workplaces; each year more than 3.3 million men and women suffer a serious job-related injury or illness. Millions more are exposed to toxic chemicals that may cause illnesses years from now.

In addition to the direct impact on individual workers, the negative consequences for America's economy are substantial. Occupational injuries and illnesses cost American employers more than $53 billion a year—more than $1 billion a week—in worker's compensation costs alone. Indirect costs to employers, including lost productivity, employee training, and replacement costs, as well as time for investigations following injuries, can more than double these costs. Workers and their families suffer great emotional and psychological costs, in addition to the loss of wages and the costs of caring for the injured, which further weakens the economy.

Application

Under the OSH law, employers are responsible for providing a safe and healthful workplace for their workers. The OSH Act covers most private sector employers

For additional information, visit qr.njatcdb.org Item #1188.

For additional information, visit qr.njatcdb.org Item #1189.

For additional information, visit qr.njatcdb.org Item #1190.

and their workers, in addition to some public sector employers and workers in the 50 states and certain territories and jurisdictions under federal authority. OSHA covers most private sector employers and their workers in all 50 states, the District of Columbia, and other U.S. jurisdictions either directly through Federal OSHA or through an OSHA-approved state plan. State plans are OSHA-approved job safety and health programs operated by individual states instead of Federal OSHA.

Enforcement

Enforcement plays and important part in OSHA's efforts to reduce workplace injuries, illnesses, and fatalities. When OSHA finds employers who fail to uphold their safety and health responsibilities, the agency takes strong, decisive actions. Inspections are initiated without advance notice, conducted using on-site or telephone and facsimile investigations, performed by highly-trained compliance officers, and scheduled based on the following priorities:

- Imminent danger
- Catastrophes—fatalities or hospitalizations
- Worker complaints or referrals
- Targeted inspections—particular hazards, high injury rates
- Follow-up inspections

Responsibility

Employers have the responsibility to provide a safe workplace. Employers must provide their workers with a workplace that does not have serious hazards and follows all OSHA safety and health standards. Employers must find and correct safety and health problems. OSHA further requires that employers must first try to eliminate or reduce hazards by making feasible changes in working conditions rather than relying on personal protective equipment such as masks, gloves, or earplugs. Switching to safer chemicals, enclosing processes to trap harmful fumes, and using ventilation systems to clean the air are examples of effective ways to eliminate or reduce risks.

THE OSH ACT AND OSHA STANDARDS AND REGULATIONS

OSHA requirements are not recommendations; rather, the requirements set forth in the OSHA standards are law. By passing the Occupational Safety and Health Act of 1970, Congress authorized enforcement of the standards developed under the Act. **See Figure 3-1.**

The OSH Act comprises 35 sections:

Section 1: Introduction
Section 2: Congressional Findings and Purpose
Section 3: Definitions
Section 4: Applicability of This Act
Section 5: Duties
Section 6: Occupational Safety and Health Standards
Section 7: Advisory Committees; Administration
Section 8: Inspections, Investigations, and Recordkeeping

Figure 3-1. THE OCCUPATIONAL SAFETY AND HEALTH ACT

Occupational Safety and Health Administration
U S Department of Labor

Occupational Safety and Health Act of 1970

To assure safe and healthful working conditions for working men and women; by authorizing enforcement of the standards developed under the Act; by assisting and encouraging the States in their efforts to assure safe and healthful working conditions; by providing for research, information, education, and training in the field of occupational safety and health; and for other purposes.

Figure 3-1. The Occupational Safety and Health Act of 1970 (OSH Act) was signed into law on December 29, 1970, establishing OSHA.

The publication *All About OSHA* also discusses standards that may be applicable in the workplace. OSHA's Construction, General Industry, Maritime, and Agriculture standards protect workers from a wide range of serious hazards. Examples of OSHA standards include requirements for employers to take the following steps:

- Provide fall protection
- Prevent trenching cave-ins
- Prevent exposure to some infectious diseases
- Ensure the safety of workers who enter confined spaces
- Prevent exposure to harmful chemicals
- Put guards on dangerous machines
- Provide respirators or other safety equipment
- Provide training for certain dangerous jobs in a language and vocabulary that workers can understand

Employer and Employee Duties

Employers must comply with the General Duty Clause of the OSH Act, which is found in Section 5, Duties. This clause requires employers to keep their workplaces free of serious recognized hazards and is generally cited when no specific OSHA standard applies to the hazard. Note also Section 5(b), which outlines employee responsibility.

Section 5, Duties

a) Each employer—
  1. shall furnish to each of his employees employment and a place of employment which are free from recognized hazards that are causing or are likely to cause death or serious physical harm to his employees;
  2. shall comply with occupational safety and health standards promulgated under this Act.

b) Each employee shall comply with occupational safety and health standards and all rules, regulations, and orders issued pursuant to this Act which are applicable to his own actions and conduct.

Standards and Regulations

OSHA's Construction, General Industry, Maritime, and Agriculture standards are parts under Title 29 of the Code of Federal Regulations (29 CFR). The subparts are generally categorized by a topic, hazard, or exposure. For example, personal protective equipment (PPE) is primarily

For additional information, visit qr.njatcdb.org Item #1191.

addressed in Subpart E of the Safety and Health Regulations for Construction (Part 1926). Many of these regulations are performance based; that is, they require something without necessarily spelling out how compliance is to be accomplished.

The two standards applicable to the majority of the work performed by Electrical Workers are found in Part 1926, Safety and Health Regulations for Construction, and in Part 1910, Occupational Safety and Health Standards. Part 1926 applies to Construction, and Part 1910 applies to General Industry.

> **1910.12(a) Standards.**
>
> The standards prescribed in part 1926 of this chapter are adopted as occupational safety and health standards under section 6 of the Act and shall apply, according to the provisions thereof, to every employment and place of employment of every employee engaged in construction work. Each employer shall protect the employment and places of employment of each of his employees engaged in construction work by complying with the appropriate standards prescribed in this paragraph.
>
> **1910.12(d)**
>
> For the purposes of this part, to the extent that it may not already be included in paragraph (b) of this section, "construction work" includes the erection of new electric transmission and distribution lines and equipment, and the alteration, conversion, and improvement of the existing transmission and distribution lines and equipment.

SAFETY AND HEALTH REGULATIONS FOR CONSTRUCTION

The 29 CFR Part 1926 Construction regulations are divided into subparts. The scope is contained in Subpart A and advises that Part 1926 sets forth the safety and health standards promulgated by the Secretary of Labor under Section 107 of the Contract Work Hours and Safety Standards Act. Currently, Part 1926 is divided into the following subparts:

1926 Subpart A:	General
1926 Subpart B:	General Interpretations
1926 Subpart C:	General Safety and Health Provisions
1926 Subpart D:	Occupational Health and Environmental Controls
1926 Subpart E:	Personal Protective and Life Saving Equipment
1926 Subpart F:	Fire Protection and Prevention
1926 Subpart G:	Signs, Signals, and Barricades
1926 Subpart H:	Materials Handling, Storage, Use, and Disposal
1926 Subpart I:	Tools—Hand and Power
1926 Subpart J:	Welding and Cutting
1926 Subpart K:	Electrical
1926 Subpart L:	Scaffolds
1926 Subpart M:	Fall Protection
1926 Subpart N:	Helicopters, Hoists, Elevators, and Conveyors
1926 Subpart O:	Motor Vehicles, Mechanized Equipment, and Marine Operations
1926 Subpart P:	Excavations
1926 Subpart Q:	Concrete and Masonry Construction
1926 Subpart R:	Steel Erection
1926 Subpart S:	Underground Construction, Caissons, Cofferdams, and Compressed Air
1926 Subpart T:	Demolition
1926 Subpart U:	Blasting and the Use of Explosives
1926 Subpart V:	Electric Power Transmission and Distribution
1926 Subpart W:	Rollover Protective Structures; Overhead Protection
1926 Subpart X:	Ladders
1926 Subpart Y:	Commercial Diving Operations
1926 Subpart Z:	Toxic and Hazardous Substances
1926 Subpart AA:	[Reserved]

For additional information, visit qr.njatcdb.org Item #1192.

1926 Subpart BB: [Reserved]
1926 Subpart CC: Cranes and
Derricks in
Construction

Note that the construction regulations are broken down into numerous subparts, each addressing a particular topic as indicated in its title and within its scope. Some subparts cover a particular topic within its scope, such as excavations, ladders, scaffolds, and welding, whereas other subparts apply more broadly, such as the general safety and health provisions of Subpart C. Some of the Part 1926 Construction regulations will be explored here as examples of the types of provisions that could apply and require compliance.

General Safety and Health Provisions

Subpart C provides general safety and health provisions as indicated by its title. Subpart C begins by providing regulations that state, in part, "no contractor or subcontractor for any part of the contract work shall require any laborer or me-

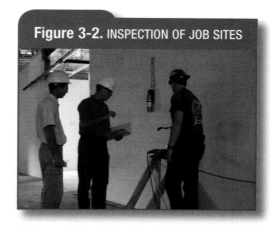

Figure 3-2. INSPECTION OF JOB SITES

Figure 3-2. *Employers must provide for frequent and regular inspections of job sites by competent persons as part of their accident prevention responsibilities.*

chanic employed in the performance of the contract to work in surroundings or under working conditions which are unsanitary, hazardous, or dangerous to his health or safety." Additional general employer provisions from Subpart C include the following points:

1926.20(b) Accident prevention responsibilities.

1926.20(b)(1)

It shall be the responsibility of the employer to initiate and maintain such programs as may be necessary to comply with this part.

1926.20(b)(2)

Such programs shall provide for frequent and regular inspections of the job sites, materials, and equipment to be made by competent persons designated by the employers. **See Figure 3-2.**

1926.20(f)(1) Personal protective equipment.

Standards in this part requiring the employer to provide personal protective equipment (PPE), including respirators and other types of PPE, because of hazards to employees impose a separate compliance duty with respect to each employee covered by the requirement. The employer must provide PPE to each employee required to use the PPE, and each failure to provide PPE to an employee may be considered a separate violation.

1926.20(f)(2) Training.

Standards in this part requiring training on hazards and related matters, such as standards requiring that employees receive training or that the employer train employees, provide training to employees, or institute or implement a training program, impose a separate compliance duty with respect to each employee covered by the requirement. The employer must train each affected employee in the manner required by the standard, and each failure to train an employee may be considered a separate violation.

1926.21(b) Employer responsibility.

1926.21(b)(2)

The employer shall instruct each employee in the recognition and avoidance of unsafe conditions and the regulations applicable to his work environment to control or eliminate any hazards or other exposure to illness or injury.

1926.28(a)

The employer is responsible for requiring the wearing of appropriate personal protective equipment in all operations where there is an exposure to hazardous conditions or where this part indicates the need for using such equipment to reduce the hazards to the employees.

Personal Protective and Life-Saving Equipment

Subpart C states that the employer is responsible for requiring employees to wear appropriate personal protective equipment in all operations where there is an exposure to hazardous conditions or where this part indicates the need for using such equipment to reduce the hazards to the employees; the employer must provide the necessary PPE to each employee who is required to use the PPE. Some of the PPE regulations from Subpart E are shown here:

1926.95(a) Application.

Protective equipment, including personal protective equipment for eyes, face, head, and extremities, protective clothing, respiratory devices, and protective shields and barriers, shall be provided, used, and maintained in a sanitary and reliable condition wherever it is necessary by reason of hazards of processes or environment, chemical hazards, radiological hazards, or mechanical irritants encountered in a manner capable of causing injury or impairment in the function of any part of the body through absorption, inhalation, or physical contact. **See Figure 3-3.**

1926.95(b) Employee-owned equipment.

Where employees provide their own protective equipment, the employer shall be responsible to assure its adequacy, including proper maintenance and sanitation of such equipment.

1926.100(b)(2)

The employer must ensure that the head protection provided for each employee exposed to high-voltage electric shock and burns also meets the specifications contained in Section 9.7 ("Electrical Insulation") of any of the consensus standards identified in paragraph (b)(1) of this section. **See Figure 3-4.** Eye and face protection.

1926.102(a) General.

1926.102(a)(1)

Employees shall be provided with eye and face protection equipment when machines or operations present potential eye or face injury from physical, chemical, or radiation agents. **See Figure 3-5.**

1926.102(a)(2)

Eye and face protection equipment required by this Part shall meet the requirements specified in American National Standards Institute, Z87.1-1968, Practice for Occupational and Educational Eye and Face Protection.

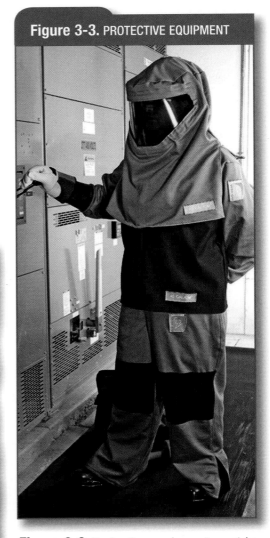

Figure 3-3. PROTECTIVE EQUIPMENT

Figure 3-3. *Protective equipment must be provided and used as necessary.* Courtesy of Salisbury by Honeywell.

OSHA Tip

Requirements for electrical protective equipment for construction are located in 1926.97. These include in-service care and use of electrical protective equipment and design requirements for specific and other types of electrical protective equipment.

Figure 3-4. PROTECTIVE HELMETS

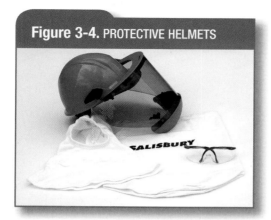

Figure 3-4. Protective helmets must be used to protect employees where there is a potential for head injury. Courtesy of Salisbury by Honeywell.

Electrical

Subpart K, Electrical, indicates that it addresses electrical safety requirements that are necessary for practical safeguarding of employees involved in construction work. This part is divided into four major divisions and applicable definitions:

1926.400(a) Installation safety requirements.

Installation safety requirements are contained in 1926.402 through 1926.408. Included in this category are electric equipment and installations used to provide electric power and light on jobsites.

1926.400(b) Safety-related work practices.

Safety-related work practices are contained in 1926.416 and 1926.417. In addition to covering the hazards arising from the use of electricity at jobsites, these regulations also cover the hazards arising from the accidental contact, direct or indirect, by employees with all energized lines, above or below ground, passing through or near the jobsite.

1926.400(c) Safety-related maintenance and environmental considerations.

Safety-related maintenance and environmental considerations are contained in 1926.431 and 1926.432.

1926.400(d) Safety requirements for special equipment. Safety requirements for special equipment are contained in 1926.441.

1926.400(e) Definitions.

Definitions applicable to this Subpart are contained in 1926.449.

Figure 3-5. EYE AND FACE PROTECTION

Figure 3-5. Eye and face protection must be provided as necessary. Courtesy of Salisbury by Honeywell.

OSHA Tip

Subpart V generally covers the construction of electric power transmission and distribution lines and equipment. Line-clearance treetrimming operations and work involving electric power generation installations must comply with 1910.269.

An overview of a number of these requirements follows:

1926.404 Wiring design and protection.

1926.404(b) Branch circuits—

1926.404(b)(1) Ground-fault protection—

1926.404(b)(1)(i) General. The employer shall use either ground fault circuit interrupters as specified in paragraph (b)(1)(ii) of this section or an assured equipment grounding conductor program assured equipment grounding conductor program as specified in paragraph (b)(1)(iii) of this section to protect employees on construction sites. These requirements are in addition to any other requirements for equipment grounding conductors.

1926.416 General requirements.

1926.416(a) Protection of employees—

1926.416(a)(1)

No employer shall permit an employee to work in such proximity to any part of an electric power circuit that the employee could contact the electric power circuit in the course of work, unless the employee is protected against electric shock by deenergizing the circuit and grounding it or by guarding it effectively by insulation or other means. **See Figure 3-6.**

1926.416(a)(2)

In work areas where the exact location of underground electric powerlines is unknown, employees using jack-hammers, bars, or other hand tools which may contact a line shall be provided with insulated protective gloves.

1926.416(a)(3)

Before work is begun the employer shall ascertain by inquiry or direct observation, or by instruments, whether any part of an energized electric power circuit, exposed or concealed, is so located that the performance of the work may bring any person, tool, or machine into physical or electrical contact with the electric power circuit. The employer shall post and maintain proper warning signs where such a circuit exists. The employer shall advise employees of the location of such lines, the hazards involved, and the protective measures to be taken.

1926.416(b) Passageways and open spaces—

1926.416(b)(1)

Barriers or other means of guarding shall be provided to ensure that workspace for electrical equipment will not be used as a passageway during periods when energized parts of electrical equipment are exposed.

1926.416(b)(2)

Working spaces, walkways, and similar locations shall be kept clear of cords so as not to create a hazard to employees.

1926.417 Lockout and tagging of circuits.

1926.417(a)

Controls. Controls that are to be deactivated during the course of work on energized or deenergized equipment or circuits shall be tagged.

1926.417(b)

Equipment and circuits. Equipment or circuits that are deenergized shall be rendered inoperative and shall have tags attached at all points where such equipment or circuits can be energized. **See Figure 3-7.**

1926.417(c)

Tags. Tags shall be placed to identify plainly the equipment or circuits being worked on.

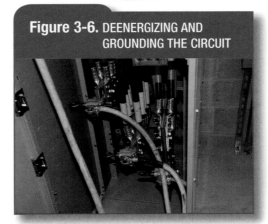

Figure 3-6. DEENERGIZING AND GROUNDING THE CIRCUIT

Figure 3-6. Deenergizing and grounding the circuit is one OSHA-recognized method of employee protection in Subpart K. Photo courtesy of National Electrical Contractors Association (NECA).

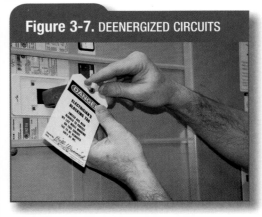

Figure 3-7. DEENERGIZED CIRCUITS

Figure 3-7. Circuits that are deenergized must be rendered inoperative and have tags attached in accordance with 1926.417(b). Photo courtesy of National Electrical Contractors Association (NECA).

OCCUPATIONAL SAFETY AND HEALTH STANDARDS

Currently, 29 CFR Part 1910 General Industry regulations are divided into 26 subparts. Note that, while a number of the subpart titles in Part 1910 are similar to those in Part 1926, they are located in a different subpart letter in each standard. For example, Fire Protection is located in Subpart F in Part 1926 and in Subpart L in Part 1910.

1910 Subpart A: General
1910 Subpart B: Adoption and Extension of Established Federal Standards
1910 Subpart C: Adoption and Extension of Established Federal Standards
1910 Subpart D: Walking–Working Surfaces
1910 Subpart E: Means of Egress
1910 Subpart F: Powered Platforms, Manlifts, and Vehicle-Mounted Work Platforms
1910 Subpart G: Occupational Health and Environmental Control
1910 Subpart H: Hazardous Materials
1910 Subpart I: Personal Protective Equipment
1910 Subpart J: General Environmental Controls
1910 Subpart K: Medical and First Aid
1910 Subpart L: Fire Protection
1910 Subpart M: Compressed Gas and Compressed Air Equipment
1910 Subpart N: Materials Handling and Storage
1910 Subpart O: Machinery and Machine Guarding
1910 Subpart P: Hand and Portable Powered Tools and Other Hand-Held Equipment
1910 Subpart Q: Welding, Cutting, and Brazing
1910 Subpart R: Special Industries
1910 Subpart S: Electrical
1910 Subpart T: Commercial Diving Operations
1910 Subpart U: [Reserved]
1910 Subpart V: [Reserved]
1910 Subpart W: Program Standard
1910 Subpart X: [Reserved]
1910 Subpart Y: [Reserved]
1910 Subpart Z: Toxic and Hazardous Substances

The requirements are generally not identical in each standard (Part 1926 and Part 1910). The electrical requirements are a good example: Compare the Subpart K electrical requirements from Part 1926 covered earlier in this chapter with the Subpart S electrical requirements from Part 1910 covered later in this chapter.

General

Subpart A of Part 1910 contains the general provisions of the Occupational Safety and Health Standards. A number of these general provisions follow. Section 1910.1 addresses the purpose and scope, Section 1910.2 contains definitions, Section 1910.6 covers incorporation by reference, and Section 1910.9 addresses compliance duties owed to each employee.

For additional information, visit qr.njatcdb.org Item #1193.

1910.1 Purpose and Scope

1910.1(a) Section 6(a) of the Williams-Steiger Occupational Safety and Health Act of 1970 (84 Stat. 1593) provides that "without regard to chapter 5 of title 5, United States Code, or to the other subsections of this section, the Secretary shall, as soon as practicable during the period beginning with the effective date of this Act and ending 2 years after such date, by rule promulgate as an occupational safety or health standard any national consensus standard, and any established Federal standard, unless he determines that the promulgation of such a standard would not result in improved safety or health for specifically designated employees." The legislative purpose of this provision is to establish, as rapidly as possible and without regard to the rule-making provisions of the Administrative Procedure Act, standards with which industries are generally familiar, and on whose adoption interested and affected persons have already had an opportunity to express their views. Such standards are either (1) national concensus standards on whose adoption affected persons have reached substantial agreement, or (2) Federal standards already established by Federal statutes or regulations.

1910.1(b) This part carries out the directive to the Secretary of Labor under section 6(a) of the Act. It contains occupational safety and health standards which have been found to be national consensus standards or established Federal standards.

1910.6(a)(1)

The standards of agencies of the U.S. Government, and organizations which are not agencies of the U.S. Government which are incorporated by reference in this part, have the same force and effect as other standards in this part. Only the mandatory provisions (i.e., provisions containing the word "shall" or other mandatory language) of standards incorporated by reference are adopted as standards under the Occupational Safety and Health Act.

1910.9(a) Personal protective equipment.

Standards in this part requiring the employer to provide personal protective equipment (PPE), including respirators and other types of PPE, because of hazards to employees impose a separate compliance duty with respect to each employee covered by the requirement. The employer must provide PPE to each employee required to use the PPE, and each failure to provide PPE to an employee may be considered a separate violation.

1910.9(b) Training.

Standards in this part requiring training on hazards and related matters, such as standards requiring that employees receive training or that the employer train employees, provide training to employees, or institute or implement a training program, impose a separate compliance duty with respect to each employee covered by the requirement. The employer must train each affected employee in the manner required by the standard, and each failure to train an employee may be considered a separate violation. **See Figure 3-8.**

Personal Protective Equipment

Personal protective equipment is covered in Subpart I of Part 1910. A number of the PPE provisions from the Occupational Safety and Health Standards are covered in this section of the text, including those related to application, design, hazard assessment and selection, training, payment, and electrical protective devices. Many of these regulations are performance based; that is, they require compliance without necessarily spelling out how to comply.

Note that the employer is required to assess the workplace to determine the need for the use of PPE. This assessment, which must be in writing, must identify the workplace evaluated, the person who performed the assessment, and the date of the assessment. PPE is also required to be provided, used, and maintained properly.

Figure 3-8. EMPLOYEE TRAINING

Figure 3-8. *The employer is required to train each affected employee in the manner required by the OSHA standard. Courtesy of Service Electric Company.*

Figure 3-9. PERSONAL PROTECTIVE EQUIPMENT

Figure 3-9. *Subpart I requires that necessary protective equipment be provided, used, and maintained in a sanitary and reliable condition. Courtesy of Salisbury by Honeywell.*

The employer is also responsible for training and retraining related to PPE, and for the adequacy of PPE, including any employee-provided PPE. Each affected employee must demonstrate an understanding of the training and the ability to use PPE properly before being allowed to perform work requiring the use of that PPE. Notice the application notes in 1910.132(g). The employer is generally required to provide PPE at no cost to employees.

1910.132(a) Application.

Protective equipment, including personal protective equipment for eyes, face, head, and extremities, protective clothing, respiratory devices, and protective shields and barriers, shall be provided, used, and maintained in a sanitary and reliable condition wherever it is necessary by reason of hazards of processes or environment, chemical hazards, radiological hazards, or mechanical irritants encountered in a manner capable of causing injury or impairment in the function of any part of the body through absorption, inhalation, or physical contact. **See Figure 3-9.**

1910.132(b) Employee-owned equipment.

Where employees provide their own protective equipment, the employer shall be responsible to assure its adequacy, including proper maintenance and sanitation of such equipment.

1910.132(d) Hazard assessment and equipment selection.

1910.132(d)(1) The employer shall assess the workplace to determine if hazards are present, or are likely to be present, which necessitate the use of personal protective equipment (PPE). If such hazards are present, or likely to be present, the employer shall:

1910.132(d)(1)(i) Select, and have each affected employee use, the types of PPE that will protect the affected employee from the hazards identified in the hazard assessment;

1910.132(d)(1)(ii) Communicate selection decisions to each affected employee; and,

1910.132(d)(1)(iii) Select PPE that properly fits each affected employee.

1910.132(d)(2) The employer shall verify that the required workplace hazard assessment has been performed through a written certification that identifies the workplace evaluated; the person certifying that the evaluation has been performed; the date(s) of the hazard assessment; and, which identifies the document as a certification of hazard assessment.

1910.132(e) Defective and damaged equipment.

Defective or damaged personal protective equipment shall not be used.

1910.132(f) Training.

1910.132(f)(1) The employer shall provide training to each employee who is required by this section to use PPE. Each such employee shall be trained to know at least the following:

1910.132(f)(1)(i) When PPE is necessary;

1910.132(f)(1)(ii) What PPE is necessary;

1910.132(f)(1)(iii) How to properly don, doff, adjust, and wear PPE;

1910.132(f)(1)(iv) The limitations of the PPE; and,

1910.132(f)(1)(v) The proper care, maintenance, useful life, and disposal of the PPE.

1910.132(f)(2)

Each affected employee shall demonstrate an understanding of the training specified in paragraph (f)(1) of this section, and the ability to use PPE properly, before being allowed to perform work requiring the use of PPE.

1910.132(f)(3)

When the employer has reason to believe that any affected employee who has already been trained does not have the understanding and skill required by paragraph (f)(2) of this section, the employer shall retrain each such employee. Circumstances where retraining is required include, but are not limited to, situations where:

1910.132(f)(3)(i) Changes in the workplace render previous training obsolete; or

1910.132(f)(3)(ii) Changes in the types of PPE to be used render previous training obsolete; or

1910.132(f)(3)(iii) Inadequacies in an affected employee's knowledge or use of assigned PPE indicate that the employee has not retained the requisite understanding or skill.

1910.132(g)

Paragraphs (d) and (f) of this section apply only to 1910.133, 1910.135, 1910.136, and 1910.138. Paragraphs (d) and (f) of this section do not apply to 1910.134 and 1910.137.

Protector gloves are generally required to be worn over insulating gloves. Insulating gloves used without protector gloves may not be reused until they have been retested. Courtesy of Salisbury by Honeywell.

Section 1910.137 addresses electrical protective equipment. These regulations cover topics such as design requirements and in-service care and use, including daily inspection and periodic electrical tests.

A number of these regulations are covered here.

Figure 3-10. MARKINGS ON GLOVES

Figure 3-10. *Markings on gloves, such as type, class, size, and manufacturer information, must be nonconductive and be confined to the cuff portion of the glove.* Courtesy of Salisbury by Honeywell.

1910.137(a)(1)(ii)

Each item shall be clearly marked as follows:

1910.137(a)(1)(ii)(A)

Class 00 equipment shall be marked Class 00.

1910.137(a)(1)(ii)(B)

Class 0 equipment shall be marked Class 0.

1910.137(a)(1)(ii)(C)

Class 1 equipment shall be marked Class 1.

1910.137(a)(1)(ii)(D)

Class 2 equipment shall be marked Class 2.

1910.137(a)(1)(ii)(E)

Class 3 equipment shall be marked Class 3.

1910.137(a)(1)(ii)(F)

Class 4 equipment shall be marked Class 4.

1910.137(a)(1)(ii)(G)

Nonozone-resistant equipment shall be marked Type I.

1910.137(a)(1)(ii)(H)

Ozone-resistant equipment shall be marked Type II.

1910.137(a)(1)(iii) Markings shall be nonconducting and shall be applied in such a manner as not to impair the insulating qualities of the equipment.

1910.137(a)(1)(iv) Markings on gloves shall be confined to the cuff portion of the glove. **See Figure 3-10.**

1910.137(c)

In-service care and use of electrical protective equipment.

1910.137(c)(1)

General. Electrical protective equipment shall be maintained in a safe, reliable condition.

1910.137(c)(2)

Specific requirements. The following specific requirements apply to rubber insulating blankets, rubber insulating covers, rubber insulating line hose, rubber insulating gloves, and rubber insulating sleeves:

1910.137(c)(2)(i)

Maximum use voltages shall conform to those listed in Table I-4. **See Figure 3-11.**

1910.137(c)(2)(ii)

Insulating equipment shall be inspected for damage before each day's use and immediately following any incident that can reasonably be suspected of causing damage. Insulating gloves shall be given an air test, along with the inspection.

Figure 3-11. TABLE I-4

TABLE I-4 Rubber Insulating Equipment, Voltage Requirements

Class of Equipment	Maximum Use Voltage[1] AC rms	Retest Voltage[2] AC rms	Retest Voltage[2] DC avg
00	500	2,500	10,000
0	1,000	5,000	20,000
1	7,500	10,000	40,000
2	17,000	20,000	50,000
3	26,500	30,000	60,000
4	36,000	40,000	70,000

[1]The maximum use voltage is the ac voltage (rms) classification of the protective equipment that designates the maximum nominal design voltage of the energized system that may be safely worked. The nominal design voltage is equal to the phase-to-phase voltage on multiphase circuits. However, the phase-to-ground potential is considered to be the nominal design voltage if: (1) There is no multiphase exposure in a system area and the voltage exposure is limited to the phase-to-ground potential, or (2) The electric equipment and devices are insulated or isolated or both so that the multiphase exposure on a grounded wye circuit is removed.

[2]The proof-test voltage shall be applied continuously for at least 1 minute, but no more than 3 minutes.

Figure 3-11. *Table I-4 provides the requirements for rubber insulation equipment voltage.*

1910.137(c)(2)(iii)

Insulating equipment with any of the following defects may not be used:

1910.137(c)(2)(iii)(A)

A hole, tear, puncture, or cut;

1910.137(c)(2)(iii)(B)

Ozone cutting or ozone checking (that is, a series of interlacing cracks produced by ozone on rubber under mechanical stress);

1910.137(c)(2)(iii)(C)

An embedded foreign object;

1910.137(c)(2)(iii)(D)

Any of the following texture changes: swelling, softening, hardening, or becoming sticky or inelastic.

1910.137(c)(2)(iii)(E)

Any other defect that damages the insulating properties.

1910.137(c)(2)(iv)

Insulating equipment found to have other defects that might affect its insulating properties shall be removed from service and returned for testing under paragraphs (c)(2) (viii) and (c)(2)(ix) of this section.

1910.137(c)(2)(v)

Insulating equipment shall be cleaned as needed to remove foreign substances

1910.137(c)(2)(vi)

Insulating equipment shall be stored in such a location and in such a manner as to protect it from light, temperature extremes, excessive humidity, ozone, and other damaging substances and conditions.

1910.137(c)(2)(vii)

Protector gloves shall be worn over insulating gloves, except as follows:

1910.137(c)(2)(vii)(A)

Protector gloves need not be used with Class 0 gloves, under limited-use conditions, when small equipment and parts manipulation necessitate unusually high finger dexterity.

Note to paragraph (c)(2)(vii)(A): Persons inspecting rubber insulating gloves used under these conditions need to take extra care in visually examining them. Employees using rubber insulating gloves under these conditions need to take extra care to avoid handling sharp objects.

1910.137(c)(2)(vii)(B)

If the voltage does not exceed 250 volts, ac, or 375 volts, dc, protector gloves need not be used with Class 00 gloves, under limited-use conditions, when small equipment and parts manipulation necessitate unusually high finger dexterity.

Note to paragraph (c)(2)(vii)(B): Persons inspecting rubber insulating gloves used under these conditions need to take extra care in visually examining them. Employees using rubber insulating gloves under these conditions need to take extra care to avoid handling sharp objects.

Figure 3-12. TABLE I-5

TABLE I-5 Rubber Insulating Equipment, Test Intervals

Type of Equipment	When to Test
Rubber insulating line hose	Upon indication that insulating value is suspect and after repair.
Rubber insulating covers	Upon indication that insulating value is suspect and after repair.
Rubber insulating blankets	Before first issue and every 12 months thereafter;[1] upon indication that insulating value is suspect; and after repair.
Rubber insulating gloves	Before first issue and every 6 months thereafter;[1] upon indication that insulating value is suspect; after repair; and after use without protectors.
Rubber insulating sleeves	Before first issue and every 12 months thereafter;[1] upon indication that insulating value is suspect; and after repair.

[1] If the insulating equipment has been electrically tested but not issued for service, the insulating equipment may not be placed into service unless it has been electrically tested within the previous 12 months.

Figure 3-12. Table I-5 details the rubber insulation equipment test intervals.

1910.137(c)(2)(vii)(C)
Any other class of glove may be used without protector gloves, under limited-use conditions, when small equipment and parts manipulation necessitate unusually high finger dexterity but only if the employer can demonstrate that the possibility of physical damage to the gloves is small and if the class of glove is one class higher than that required for the voltage involved.

1910.137(c)(2)(vii)(D)
Insulating gloves that have been used without protector gloves may not be reused until they have been tested under the provisions of paragraphs (c)(2)(viii) and (c)(2)(ix) of this section.

1910.137(c)(2)(viii)
Electrical protective equipment shall be subjected to periodic electrical tests. Test voltages and the maximum intervals between tests shall be in accordance with Table I-4 and Table I-5. **See Figure 3-12.**

The Control of Hazardous Energy (Lockout/Tagout)

Lockout/tagout regulations are primarily addressed in two subparts in the Part 1910, Occupational Safety and Health Standards: Subpart J, General Environmental Controls, and Subpart S, Electrical.

OSHA Tip

Section 1910.269 covers the operation and maintenance of electric power generation, control, transformation, transmission, and distribution lines and equipment.

The Subpart J regulations are located in 1910.147, The Control of Hazardous Energy (lockout/tagout). Section 1910.147 is organized into the following subdivisions:

- Scope, application, and purpose
- Definitions
- General
- Application of control
- Lockout or tagout devices removal
- Additional requirements

The Subpart S lockout/tagout regulations are located in 1910.333(b)(2), Lockout and Tagging. One section and an accompanying note from these Subpart S requirements are included here for reference. These lockout and tagging regulations must be reviewed in their entirety for a complete look at what is required. The provisions of 1910.333(b)(2) are

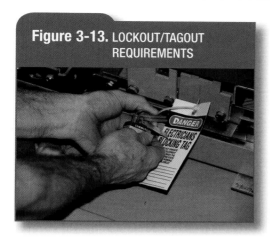

Figure 3-13. LOCKOUT/TAGOUT REQUIREMENTS

Figure 3-13. Both a lock and a tag are generally required to comply with lockout/ tagout requirements while any employee is exposed to contact with parts of fixed electric equipment or circuits that have been deenergized. Photo courtesy of National Electrical Contractors Association (NECA).

dresses electrical safety requirements that are necessary for the practical safeguarding of employees in their workplaces and is divided into four major divisions:

- Design safety standards for electrical systems
- Safety-related work practices
- Safety-related maintenance requirements
- Safety requirements for special equipment

The primary focus is on a number of the safety-related work practice regulations of Subpart S, including some from 1910.332, Training; 1910.333, Selection and Use of Work Practices; 1910.334, Use of Equipment; and 1910.335, Safeguards for Personnel Protection. In addition, a number of the definitions applicable to Subpart S located in 1910.399 are included here for review purposes.

covered in greater detail in the chapter, "Introduction to Lockout, Tagging, and Control of Hazardous Energy."

1910.333(b)(2) Lockout and Tagging.

While any employee is exposed to contact with parts of fixed electric equipment or circuits which have been deenergized, the circuits energizing the parts shall be locked out or tagged or both in accordance with the requirements of this paragraph. **See Figure 3-13.** The requirements shall be followed in the order in which they are presented.

Note 2: Lockout and tagging procedures that comply with paragraphs (c) through (f) of 1910.147 will also be deemed to comply with paragraph (b)(2) of this section provided that:

[1] The procedures address the electrical safety hazards covered by this Subpart; and

[2] The procedures also incorporate the requirements of paragraphs (b)(2)(iii)(D) and (b)(2)(iv)(B) of this section.

Electrical Occupational Safety and Health Standards

Subpart S details the electrical regulations of Part 1910, Occupational Safety and Health Standards. This subpart ad-

1910.332(b) Content of training.

1910.332(b)(1) Practices addressed in this standard.

Employees shall be trained in and familiar with the safety-related work practices required by 1910.331 through 1910.335 that pertain to their respective job assignments.

1910.333(a) General.

Safety-related work practices shall be employed to prevent electric shock or other injuries resulting from either direct or indirect electrical contacts, when work is performed near or on equipment or circuits which are or may be energized. The specific safety-related work practices shall be consistent with the nature and extent of the associated electrical hazards.

1910.333(a)(1) Deenergized parts.

Live parts to which an employee may be exposed shall be deenergized before the employee works on or near them, unless the employer can demonstrate that deenergizing introduces additional or increased hazards or is infeasible due to equipment design or operational limitations. Live parts that operate at less than 50 volts to ground need not be deenergized if there will be no increased exposure to electrical burns or to explosion due to electric arcs.

Note 1: Examples of increased or additional hazards include interruption of life support equipment, deactivation of emergency alarm systems, shutdown of hazardous location ventilation equipment, or removal of illumination for an area.

Note 2: Examples of work that may be performed on or near energized circuit parts because of infeasibility due to equipment design or operational limitations include testing of electric circuits that can only be performed with the circuit energized and work on circuits that form an integral part of a continuous industrial process in a chemical plant that would otherwise need to be completely shut down in order to permit work on one circuit or piece of equipment.

Note 3: Work on or near deenergized parts is covered by paragraph (b) of this section.

1910.333(a)(2) Energized parts.

If the exposed live parts are not deenergized (i.e., for reasons of increased or additional hazards or infeasibility), other safety-related work practices shall be used to protect employees who may be exposed to the electrical hazards involved. **See Figure 3-14.** Such work practices shall protect employees against contact with energized circuit parts directly with any part of their body or indirectly through some other conductive object. The work practices that are used shall be suitable for the conditions under which the work is to be performed and for the voltage level of the exposed electric conductors or circuit parts. Specific work practice requirements are detailed in paragraph (c) of this section.

1910.333(b) Working on or near exposed deenergized parts.

1910.333(b)(1) Application.

This paragraph applies to work on exposed deenergized parts or near enough to them to expose the employee to any electrical hazard they present. Conductors and parts of electric equipment that have been deenergized but have not been locked out or tagged in accordance with paragraph (b) of this section shall be treated as energized parts, and paragraph (c) of this section applies to work on or near them.

1910.333(c) Working on or near exposed energized parts.

1910.333(c)(1)

"Application." This paragraph applies to work performed on exposed live parts (involving either direct contact or by means of tools or materials) or near enough to them for employees to be exposed to any hazard they present.

1910.333(c)(2) Work on energized equipment.

Only qualified persons may work on electric circuit parts or equipment that have not been deenergized under the procedures of paragraph (b) of this section. Such persons shall be capable of working safely on energized circuits and shall be familiar with the proper use of special precautionary techniques, personal protective equipment, insulating and shielding materials, and insulated tools.

1910.333(c)(3) Overhead lines.

If work is to be performed near overhead lines, the lines shall be deenergized and grounded, or other protective measures shall be provided before work is started. **See Figure 3-15.** If the lines are to be deenergized, arrangements shall be made with the person or organization that operates or controls the electric circuits involved to deenergize and ground them. If protective measures, such as guarding, isolating, or insulating, are provided, these precautions shall prevent employees from contacting such lines directly with any part of their body or indirectly through conductive materials, tools, or equipment.

Note: The work practices used by qualified persons installing insulating devices on overhead power transmission or distribution lines are covered by 1910.269 of this Part, not by 1910.332 through 1910.335 of this Part. Under paragraph (c)(2) of this section, unqualified persons are prohibited from performing this type of work.

1910.333(c)(3)(i) Unqualified persons.

1910.333(c)(3)(i)(A)

When an unqualified person is working in an elevated position near overhead lines, the location shall be such that the person and the longest conductive object he or she may contact cannot come closer to any unguarded, energized overhead line than the following distances:

1910.333(c)(3)(i)(A)(1)

For voltages to ground 50kV or below—10 feet (305 cm);

1910.333(c)(3)(i)(A)(2)

For voltages to ground over 50kV—10 feet (305 cm) plus 4 inches (10 cm) for every 10kV over 50kV.

Figure 3-14. EXPOSURE TO ELECTRICAL HAZARDS

Figure 3-14. *Other safety-related work practices must be used to protect employees who may be exposed to the electrical hazards involved if the exposed live parts are not deenergized.* Courtesy of Salisbury by Honeywell.

Figure 3-15. OVERHEAD LINES

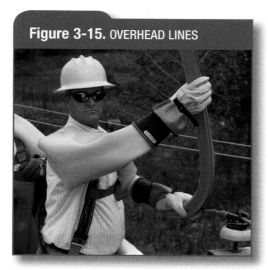

Figure 3-15. *Other protective measures must be provided before work is started if that work is to be performed near overhead lines and the lines are not deenergized and grounded.* Courtesy of Salisbury by Honeywell.

1910.333(c)(3)(i)(B)

When an unqualified person is working on the ground in the vicinity of overhead lines, the person may not bring any conductive object closer to unguarded, energized overhead lines than the distances given in paragraph (c)(3)(i)(A) of this section.

Note: For voltages normally encountered with overhead power line, objects which do not have an insulating rating for the voltage involved are considered to be conductive.

1910.333(c)(3)(ii) Qualified persons.

When a qualified person is working in the vicinity of overhead lines, whether in an elevated position or on the ground, the person may not approach or take any conductive object without an approved insulating handle closer to exposed energized parts than shown in Table S-5 unless:

1910.333(c)(3)(ii)(A)

The person is insulated from the energized part (gloves, with sleeves if necessary, rated for the voltage involved are considered to be insulation of the person from the energized part on which work is performed), or

1910.333(c)(3)(ii)(B)

The energized part is insulated both from all other conductive objects at a different potential and from the person, or

1910.333(c)(3)(ii)(C)

The person is insulated from all conductive objects at a potential different from that of the energized part.

1910.333(c)(3)(iii) Vehicular and mechanical equipment.

1910.333(c)(3)(iii)(A)

Any vehicle or mechanical equipment capable of having parts of its structure elevated near energized overhead lines shall be operated so that a clearance of 10 ft. (305 cm) is maintained. If the voltage is higher than 50kV, the clearance shall be increased 4 in. (10 cm) for every 10kV over that voltage. However, under any of the following conditions, the clearance may be reduced:

1910.333(c)(3)(iii)(A)(1)

If the vehicle is in transit with its structure lowered, the clearance may be reduced to 4 ft. (122 cm). If the voltage is higher than 50kV, the clearance shall be increased 4 in. (10 cm) for every 10 kV over that voltage.

1910.333(c)(3)(iii)(A)(2)

If insulating barriers are installed to prevent contact with the lines, and if the barriers are rated for the voltage of the line being guarded and are not a part of or an attachment to the vehicle or its raised structure, the clearance may be reduced to a distance within the designed working dimensions of the insulating barrier.

1910.333(c)(3)(iii)(A)(3)

If the equipment is an aerial lift insulated for the voltage involved, and if the work is performed by a qualified person, the clearance (between the uninsulated portion of the aerial lift and the power line) may be reduced to the distance given in Table S-5. **See Figure 3-16.**

1910.333(c)(3)(iii)(B)

Employees standing on the ground may not contact the vehicle or mechanical equipment or any of its attachments, unless:

1910.333(c)(3)(iii)(B)(1)

The employee is using protective equipment rated for the voltage; or

1910.333(c)(3)(iii)(B)(2)

The equipment is located so that no uninsulated part of its structure (that portion of the structure that provides a conductive path to employees on the ground) can come closer to the line than permitted in paragraph (c)(3)(iii) of this section.

1910.333(c)(3)(iii)(C)

If any vehicle or mechanical equipment capable of having parts of its structure elevated near energized overhead lines is intentionally grounded, employees working on the ground near the point of grounding may not stand at the grounding location whenever there is a possibility of overhead line contact. Additional precautions, such as the use of barricades or insulation, shall be taken to protect employees from hazardous ground potentials, depending on earth resistivity and fault currents, which can develop within the first few feet or more outward from the grounding point.

1910.333(c)(4) Illumination.

1910.333(c)(4)(i)

Employees may not enter spaces containing exposed energized parts, unless illumination is provided that enables the employees to perform the work safely.

1910.333(c)(4)(ii)

Where lack of illumination or an obstruction precludes observation of the work to be performed, employees may not perform tasks near exposed energized parts. Employees may not reach blindly into areas which may contain energized parts.

1910.333(c)(5) Confined or enclosed work spaces.

When an employee works in a confined or enclosed space (such as a manhole or vault) that contains exposed energized parts, the employer shall provide, and the employee shall use, protective shields, protective barriers, or insulating materials as necessary to avoid inadvertent contact with these parts. Doors, hinged panels, and the like shall be secured to prevent their swinging into an employee and causing the employee to contact exposed energized parts.

1910.333(c)(6) Conductive materials and equipment.

Figure 3-16. MINIMUM APPROACH DISTANCES

OSHA 1910.333 Table S-5 Approach Distances for Qualified Employees: Alternating Current

Voltage Range (Phase to Phase)	Minimum Approach Distance
300 V and less	Avoid Contact
Over 300 V, not over 750 V	1 ft. 0 in. (30.5 cm)
Over 750 V, not over 2 kV	1 ft. 6 in. (46 cm)
Over 2 kV, not over 15 kV	2 ft. 0 in. (61 cm)
Over 15 kV, not over 37 kV	3 ft. 0 in. (91 cm)
Over 37kV, not over 87.5 kV	3 ft. 6 in. (107 cm)
Over 87.5 kV, not over 121 kV	4 ft. 0 in. (122 cm)
Over 121 kV, not over 140 kV	4 ft. 6 in. (137 cm)

Figure 3-16. The minimum approach distances for alternating current for qualified employees are set forth in Table S-5 in Subpart S of Part 1910.

Conductive materials and equipment that are in contact with any part of an employee's body shall be handled in a manner that will prevent them from contacting exposed energized conductors or circuit parts. If an employee must handle long dimensional conductive objects (such as ducts and pipes) in areas with exposed live parts, the employer shall institute work practices (such as the use of insulation, guarding, and material handling techniques) which will minimize the hazard.

1910.333(c)(7) Portable ladders.

Portable ladders shall have nonconductive siderails if they are used where the employee or the ladder could contact exposed energized parts.

1910.333(c)(8) Conductive apparel.

Conductive articles of jewelry and clothing (such as watch bands, bracelets, rings, key chains, necklaces, metalized aprons, cloth with conductive thread, or metal headgear) may not be worn if they might contact exposed energized parts. However, such articles may be worn if they are rendered nonconductive by covering, wrapping, or other insulating means.

1910.333(c)(9) Housekeeping duties.

Where live parts present an electrical contact hazard, employees may not perform housekeeping duties at such close distances to the parts that there is a possibility of contact, unless adequate safeguards (such as insulating equipment or barriers) are provided. Electrically conductive cleaning materials (including conductive solids such as steel wool, metalized cloth, and silicon carbide, as well as conductive liquid solutions) may not be used in proximity to energized parts unless procedures are followed which will prevent electrical contact.

1910.333(c)(10) Interlocks.

Only a qualified person following the requirements of paragraph (c) of this section may defeat an electrical safety interlock, and then only temporarily while he or she is working on the equipment. The interlock system shall be returned to its operable condition when this work is completed.

1910.334(b)(2) Reclosing circuits after protective device operation.

After a circuit is deenergized by a circuit protective device, the circuit may not be manually reenergized until it has been determined that the equipment and circuit can be safely energized. The repetitive manual reclosing of circuit breakers or reenergizing circuits through replaced fuses is prohibited.

Note: When it can be determined from the design of the circuit and the overcurrent devices involved that the automatic operation of a device was caused by an overload rather than a fault condition, no examination of the circuit or connected equipment is needed before the circuit is reenergized.

1910.335(a) Use of protective equipment.

1910.335(a)(1) Personal protective equipment.

1910.335(a)(1)(i)

Employees working in areas where there are potential electrical hazards shall be provided with, and shall use, electrical protective equipment that is appropriate for the specific parts of the body to be protected and for the work to be performed.

Note: Personal protective equipment requirements are contained in subpart I of this part.

1910.335(a)(1)(ii)

Protective equipment shall be maintained in a safe, reliable condition and shall be periodically inspected or tested, as required by 1910.137.

1910.335(a)(1)(iii)

If the insulating capability of protective equipment may be subject to damage during use, the insulating material shall be protected. (For example, an outer covering of leather is sometimes used for the protection of rubber insulating material.)

1910.335(a)(1)(iv)

Employees shall wear nonconductive head protection wherever there is a danger of head injury from electric shock or burns due to contact with exposed energized parts.

1910.335(a)(1)(v)

Employees shall wear protective equipment for the eyes or face wherever there is danger of injury to the eyes or face from electric arcs or flashes or from flying objects resulting from electrical explosion.

1910.335(a)(2) General protective equipment and tools.

1910.335(a)(2)(i)

When working near exposed energized conductors or circuit parts, each employee shall use insulated tools or handling equipment if the tools or handling equipment might make contact with such conductors or parts. **See Figure 3-17.** If the insulating capability of insulated tools or handling equipment is subject to damage, the insulating material shall be protected.

1910.335(a)(2)(i)(A)

Fuse handling equipment, insulated for the circuit voltage, shall be used to remove or install fuses when the fuse terminals are energized.

1910.335(a)(2)(i)(B)

Ropes and handlines used near exposed energized parts shall be nonconductive.

1910.335(a)(2)(ii)

Protective shields, protective barriers, or insulating materials shall be used to protect each employee from shock, burns, or other electrically related injuries while that employee is working near exposed energized parts which might be accidentally contacted or where dangerous electric heating or arcing might occur. **See Figure 3-18.** When normally enclosed live parts are exposed for maintenance or repair, they shall be guarded to protect unqualified persons from contact with the live parts.

1910.335(b) Alerting techniques.

The following alerting techniques shall be used to warn and protect employees from hazards which could cause injury due to electric shock, burns, or failure of electric equipment parts:

1910.335(b)(1) Safety signs and tags.

Safety signs, safety symbols, or accident prevention tags shall be used where necessary to warn employees about electrical hazards which may endanger them, as required by 1910.145.

1910.335(b)(2) Barricades.

Barricades shall be used in conjunction with safety signs where it is necessary to pre-

Figure 3-17. INSULATED TOOLS

Figure 3-17. OSHA requires that insulated tools be used if the tool could contact exposed energized conductors or circuit parts when employees are working on or near such conductors or parts. Courtesy of Klein Tools.

Figure 3-18. ARC SUPPRESSION BLANKET

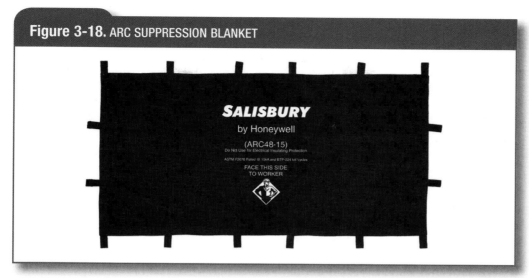

Figure 3-18. An arc suppression blanket is not designed for electrical insulating protection.
Courtesy of Salisbury by Honeywell.

vent or limit employee access to work areas exposing employees to uninsulated energized conductors or circuit parts. Conductive barricades may not be used where they might cause an electrical contact hazard.

1910.335(b)(3) Attendants.

If signs and barricades do not provide sufficient warning and protection from electrical hazards, an attendant shall be stationed to warn and protect employees.

1910.399, Definitions, in part.

Deenergized. Free from any electrical connection to a source of potential difference and from electrical charge; not having a potential different from that of the earth.

Energized. Electrically connected to a source of potential difference.

Overcurrent. Any current in excess of the rated current of equipment or the ampacity of a conductor. It may result from overload, short circuit, or ground fault.

Overload. Operation of equipment in excess of normal, full-load rating, or of a conductor in excess of rated ampacity that, when it persists for a sufficient length of time, would cause damage or dangerous overheating. A fault, such as a short circuit or ground fault, is not an overload. (See Overcurrent.)

Qualified person. One who has received training in and has demonstrated skills and knowledge in the construction and operation of electric equipment and installations and the hazards involved.

Note 1 to the definition of "qualified person:" Whether an employee is considered to be a "qualified person" will depend upon various circumstances in the workplace. For example, it is possible and, in fact, likely for an individual to be considered "qualified" with regard to certain equipment in the workplace, but "unqualified" as to other equipment. (See 1910.332(b)(3) for training requirements that specifically apply to qualified persons.)

Note 2 to the definition of "qualified person:" An employee who is undergoing on-the-job training and who, in the course of such training, has demonstrated an ability to perform duties safely at his or her level of training and who is under the direct supervision of a qualified person is considered to be a qualified person for the performance of those duties.

Summary

Every day, workers are exposed to hazards associated with working on energized parts, or hazards associated with working close enough to energized parts to allow potential exposure to electrical hazards. When requirements are adhered to and appropriate procedures are in place and implemented, the likelihood of adverse incidents, injuries, and fatalities can be reduced. Many of these requirements were discussed in this chapter. However, the requirements covered here are only a few of the full set that must be considered when assessing workplace hazards and developing an effective safety program.

An understanding of the General Duty Clause; the Part 1926, Safety and Health Regulations for Construction; and the Part 1910, Occupational Safety and Health Standards, is important. Implementing and following the requirements, such as those for training, selection and use of work practices, the use of equipment, and safeguards for personnel protection, will help ensure that the workplace is free of recognized hazards.

This knowledge will also provide a foundation for what is required by OSHA and for how the requirements of *NFPA 70E: Standard for Electrical Safety in the Workplace*® may be an aid in accomplishing what OSHA requires. Consider this analogy: "OSHA is the shall, and NFPA 70E is the how" as you explore the requirements of *NFPA 70E*. Determine whether the provisions of *NFPA 70E* can be a compliance solution and what, if anything, *NFPA 70E* requires that OSHA does not already require.

Review Questions

1. The Occupational Safety and Health Act (OSH Act) of 1970 states that its purpose is to assure so far as possible every working man and woman in the nation safe and __?__ working conditions and to preserve our human resources.
   a. convenient
   b. healthful
   c. legal
   d. reliable

2. The primary focus of this chapter are the regulations in Part __?__, Safety and Health Regulations for Construction, and Part __?__, General Industry Occupational Safety and Health Standards.
   a. 1910/1926
   b. 1926/1910
   c. 1970/1972
   d. 1992/2022

3. OSHA's safety and health standards have prevented countless work-related injuries, illnesses, and deaths. Unfortunately, far too many __?__ injuries and fatalities continue to occur.
   a. obvious
   b. preventable
   c. unavoidable
   d. unpreventable

4. Under the OSHA __?__, employers are responsible for providing a safe and healthful workplace for their workers.
   a. guidelines
   b. law
   c. recommendations
   d. wishes

5. OSHA's rules and regulations are __?__, they are not recommendations.
   a. guidelines
   b. non-mandatory
   c. requirements
   d. wishes

6. Many of the Occupational Safety and Health Standards regulations are __?__-based, which means that they require something without necessarily spelling out how compliance is to be accomplished.
   a. common sense
   b. fact
   c. performance
   d. prescriptive

7. The employer must provide for frequent and regular inspections of job sites by __?__ persons as part of their accident prevention responsibilities.
   a. competent
   b. independent
   c. qualified
   d. recognized

8. Subpart C requires that the __?__ is responsible for employees wearing appropriate personal protective equipment in all operations where there is an exposure to hazardous conditions.
   a. employee
   b. employer
   c. union
   d. warehouse supervisor

9. OSHA calls for personal protective equipment (PPE) to be provided by the __?__, which is required to assess the workplace to determine the need for the use of PPE.
   a. employee
   b. employer
   c. union
   d. warehouse supervisor

10. Each day workers are exposed to hazards, but when __?__ are adhered to and appropriate procedures are in place and implemented, incidents, injuries, and fatalities can be reduced.
    a. guidelines
    b. recommendations
    c. requirements
    d. wishes

Introduction to Lockout, Tagging, and the Control of Hazardous Energy

Chapter Outline

- The Occupational Safety and Health Act of 1970
- Safety and Health Regulations for Construction
- Occupational Safety and Health Standards
- Achieving an Electrically Safe Work Condition

Chapter Objectives

1. Demonstrate an understanding of the reasons why the Occupational Safety and Health Act of 1970 was enacted, the Congressional finding and purpose of the act, and the employer and employee responsibilities established in the General Duty Clause.
2. Become familiar with Safety and Health Regulations for Construction, including Subpart C, General Safety and Health Provisions, and the Subpart K provisions for lockout and tagging of circuits.
3. Demonstrate an understanding of the Occupational Safety and Health Standards, including the Subpart S provisions related to lockout and tagging, and the six major headings of the control of hazardous energy (lockout/tagout) in Subpart J.
4. Discuss the structure of Article 120 and the six-step process required to verify an electrically safe work condition.

Chapter 4

References

1. *NFPA 70E®*, 2015 Edition
2. OSHA 29 CFR Part 1910
3. OSHA 29 CFR Part 1926
4. The Occupational Safety and Health Act of 1970

Case Study

A 48-year-old machine operator was killed when he was crushed inside a machine. The employer had been in business for more than 80 years and had approximately 175 employees. There were 55 employees at the facility where the incident occurred. The victim had been employed with the company for 19 years.

The employer had a written safety program with task-specific safe work procedures for all positions in the shop. Employees held informal weekly tailgate safety meetings with the supervisors, as well as formal monthly safety meetings. The company's training program was usually accomplished through on-the-job-training monitored by the supervisors.

The machine involved in the incident was completely automated. The operating portion of the machine was enclosed for safety. Whenever any regular access panel or door to the machine was opened, the machine was supposed to shut off automatically. The pedestal on which the machine sat was also enclosed, with the exception of the area where the conveyor belt exited from underneath the machine. The guarding around the pedestal had to be mechanically removed to gain access to the pedestal.

On the day of the incident, the machine stopped working, and its warning lights came on. The victim was not at his workstation. His coworkers contacted the supervisor, who discovered the victim inside the machine with his head trapped between the pedestal frame and the mobile plate frame. Coworkers called 911, and the responding fire fighters had to unbolt the guarding around the pedestal frame to gain access to the victim and extricate him. Paramedics pronounced the victim dead after they removed him from the machine.

Investigation of the incident site revealed the victim's tools lying next to the opening in the pedestal where the conveyor belt was located; in addition, several pieces of the machine product were scattered about the conveyor belt. The guard or shield on the side of the conveyor belt was unbolted on one side. These factors suggested that the victim crawled into the machine from the opening for the conveyor belt.

The cause of death, according to the death certificate, was blunt head trauma.

Source: For details of this case, see FACE Investigation #03CA006. Accessed June 4, 2012.

For additional information, visit qr.njatcdb.org Item #1194.

INTRODUCTION

The Occupational Safety and Health Administration's (OSHA's) requirements are performance oriented in many cases; that is, protection of workers is required, although the requirements do not necessarily spell out precisely how worker protection is to be accomplished. This includes requirements related to protecting workers from electrical hazards. While increasingly more employers are looking to *NFPA 70E* for guidance in an effort to understand how to comply with OSHA's performance-oriented requirements, OSHA's requirements related to lockout/tagout and working on or near exposed deenergized parts are examples where OSHA offers more specific details, including the steps that must be followed to comply with these requirements.

This chapter explores OSHA's requirements for lockout and tagging of circuits [1926.417], control of hazardous energy (lockout/tagout) [1910.147], the lockout and tagging provisions of 1910.333(b)(2), and the ways that provisions in *NFPA 70E* Article 120 can meet or supplement OSHA's provisions, including those in the Occupational Safety and Health Act of 1970.

Many lockout/tagout programs are based on OSHA 1910.147. OSHA notes that lockout and tagging procedures complying with 1910.147 are also deemed to comply with 1910.333(b)(2), provided that the procedures address the electrical safety hazards covered by Subpart S of 29 CFR 1910, and that those procedures incorporate the requirements of two additional paragraphs not contained in 1910.147.

In addition to developing lockout/tagout programs based on requirements from OSHA, the requirements from *NFPA 70E* Article 120, Establishing an Electrically Safe Work Condition, should be considered as part of electrical safety practices. The provisions of *NFPA 70E* Article 120 generally meet or exceed OSHA's lockout/ tagout requirements and should be examined carefully for any other considerations that might lead to a more comprehensive lockout/tagout program. This includes, but is not limited to, the six-step verification process outlined in Section 120.1 to verify that an electrically safe work condition exists after the provisions contained in *NFPA 70E* Section 120.2 have been taken into account. The provisions of Article 120 need to be reviewed in their entirety for a full understanding of their use and application.

THE OCCUPATIONAL SAFETY AND HEALTH ACT OF 1970

The Occupational Safety and Health Act of 1970 was enacted "to assure safe and healthful working conditions for working men and women; by authorizing enforcement of the standards developed under the Act; by assisting and encouraging the States in their efforts to assure safe and healthful working conditions; by providing for research, information, education, and training in the field of occupational safety and health; and for other purposes." This Act consists of 35 sections. Part of the content of two of those sections follows:

Section 2, Congressional Findings and Purpose

(a) The Congress finds that personal injuries and illnesses arising out of work situations impose a substantial burden upon, and are a hindrance to, interstate commerce in terms of lost production, wage loss, medical expenses, and disability compensation payments, and that

(b) The Congress declares it to be its purpose and policy, through the exercise of its powers to regulate commerce among the several States and with foreign nations and to provide for the general welfare, to assure so far as possible every working man and woman in the Nation safe and healthful working conditions and to preserve our human resources.

Section 2(b) includes 13 points that detail how to accomplish the purpose and satisfy the policy of assuring, so far

OSHA Tip

The standards contained in 29 CFR 1926 Subpart V apply to the construction of electric power transmission and distribution lines and equipment.

as possible, that every working man and woman in the nation has safe and healthful working conditions and of preserving human resources.

Section 5, Duties, is commonly known as "the General Duty Clause." It identifies responsibilities for both employers and employees:

> (a) Each employer—
>   (1) shall furnish to each of his employees employment and a place of employment which are free from recognized hazards that are causing or are likely to cause death or serious physical harm to his employees;
>   (2) shall comply with occupational safety and health standards promulgated under this Act.
> (b) Each employee shall comply with occupational safety and health standards and all rules, regulations, and orders issued pursuant to this Act which are applicable to his own actions and conduct.

A fundamental understanding of the reason why the Occupational Safety and Health Act of 1970 was enacted, the Congressional finding and purpose in enacting this legislation, and employer and employee responsibilities is necessary as a foundation for understanding why safe work practices are essential "to assure safe and healthful working conditions for working men and women." Lockout and tagging of circuits, control of hazardous energy, and achieving an electrically safe work condition are among the safe work practices intended to accomplish this goal.

SAFETY AND HEALTH REGULATIONS FOR CONSTRUCTION

Part 1926, Safety and Health Regulations for Construction, sets forth the safety and health standards promulgated by the Secretary of Labor under Section 107 of the Contract Work Hours and Safety Standards Act. Subpart C sets forth the general safety and health regulations of Part 1926, while Subpart K addresses electrical safety requirements within the scope of that subpart.

General Safety and Health Provisions

Subpart C provides general safety and health provisions for Part 1926, Safety and Health Regulations for Construction. The following definition and general provisions help clarify the application of Part 1926. Subpart C is to be referred to in its entirety for a complete understanding of these regulations.

Figure 4-1. EMPLOYEE INSTRUCTION

Figure 4-1. *OSHA's general safety and health provisions require that employees be instructed in any regulations applicable to their work and be able to recognize and avoid unsafe conditions.* Courtesy of Ideal Industries, Inc.

1926.21(b)(2)

The employer shall instruct each employee in the recognition and avoidance of unsafe conditions and the regulations applicable to his work environment to control or eliminate any hazards or other exposure to illness or injury. **See Figure 4-1**.

1926.20(f)(2)

Standards in this part requiring training on hazards and related matters, such as standards requiring that employees receive training or that the employer train employees, provide training to employees, or institute or implement a training program, impose a separate compliance duty with respect to each employee covered by the requirement. The employer must train each affected employee in the manner required by the standard, and each failure to train an employee may be considered a separate violation.

1926.32(g)

For purposes of this section, "Construction work" means work for construction, alteration, and/or repair, including painting and decorating.

Electrical Safety Requirements

Subpart K, Electrical, addresses electrical safety requirements that are necessary for the practical safeguarding of employees involved in construction work. Subpart K is divided into four major divisions plus an applicable definitions section. Safety-related work practices—one of those four major divisions—are contained in Sections 1926.416 and 1926.417, with Section 1926.417 addressing lockout and tagging of circuits.

1926.417 Lockout and tagging of circuits.
1926.417(a)

Controls. Controls that are to be deactivated during the course of work on energized or deenergized equipment or circuits shall be tagged.

1926.417(b)

Equipment and circuits. Equipment or circuits that are deenergized shall be rendered inoperative and shall have tags attached at all points where such equipment or circuits can be energized. **See Figure 4-2**.

1926.417(c)

Tags. Tags shall be placed to identify plainly the equipment or circuits being worked on.

Figure 4-2. DEENERGIZED EQUIPMENT AND CIRCUITS

Figure 4-2. Deenergized equipment or circuits shall be rendered inoperative and have tags attached at all points where such equipment or circuits can be energized. The tags must be placed so as to clearly identify the equipment or circuits being worked on.
Courtesy of Ideal Industries, Inc.

OCCUPATIONAL SAFETY AND HEALTH STANDARDS

The title of Part 1910 is the Occupational Safety and Health Standards. Per 1910.1(b), Part 1910 carries out the directive of the Secretary of Labor under section 6(a) of the Occupational Safety and Health Act of 1970. It contains occupational safety and health standards that have been found to be national consensus standards or established federal standards. A number of provisions from Subparts J and S are examined here. Subpart S, Electrical, includes 1910.333(b)(2), Lockout and Tagging. Subpart J, General Environmental Controls, includes 1910.147, The control of hazardous energy (lockout/tagout).

Lockout and Tagging

Part 1910, Subpart S, addresses electrical safety requirements that are necessary for the practical safeguarding of employees in their workplaces. It is divided into four major divisions, one of which contains safety-related work practice regulations, in Sections 1910.331 through 1910.360. Lockout and tagging within the scope of Subpart S is covered within 1910.333(b), which deals with working on or near exposed deenergized parts. **See Figure 4-3**.

Figure 4-3. LOCKS AND TAGS

Figure 4-3. Both locks and tags are generally required when working on or near exposed deenergized parts. Courtesy of Ideal Industries, Inc.

Figure 4-4. LOCKOUT/TAGOUT DEVICES

Figure 4-4. Appropriate lockout and tagout devices must be used in compliance with any applicable requirements while any employee is exposed to contact with equipment or circuits that have been deenergized. Courtesy of Ideal Industries, Inc.

1910.333(b)(1) Application.
This paragraph applies to work on exposed deenergized parts or near enough to them to expose the employee to any electrical hazard they present. Conductors and parts of electric equipment that have been deenergized but have not been locked out or tagged in accordance with paragraph (b) of this section shall be treated as energized parts, and paragraph (c) of this section applies to work on or near them.

1910.333(b)(2) Lockout and Tagging.
While any employee is exposed to contact with parts of fixed electric equipment or circuits which have been deenergized, the circuits energizing the parts shall be locked out or tagged or both in accordance with the requirements of this paragraph. **See Figure 4-4**.

The requirements shall be followed in the order in which they are presented (i.e., paragraph (b)(2)(i) first, then paragraph (b)(2)(ii), etc.).

Note 1: As used in this section, fixed equipment refers to equipment fastened in place or connected by permanent wiring methods.

Note 2: Lockout and tagging procedures that comply with paragraphs (c) through (f) of 1910.147 will also be deemed to comply with paragraph (b)(2) of this section provided that:

[1] The procedures address the electrical safety hazards covered by this Subpart; and

[2] The procedures also incorporate the requirements of paragraphs (b)(2)(iii)(D) and (b)(2)(iv)(B) of this section

Procedures

1910.333(b)(2)(i)
The employer shall maintain a written copy of the procedures outlined in paragraph (b)(2) and shall make it available for inspection by employees and by the Assistant Secretary of Labor and his or her authorized representatives. **See Figure 4-5**.

Note: The written procedures may be in the form of a copy of paragraph (b) of this section.

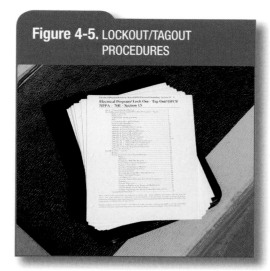

Figure 4-5. LOCKOUT/TAGOUT PROCEDURES

Figure 4-5. The lockout and tagging proce-dures must be in writing and made available for inspection. Photo courtesy of National Electrical Contractors Association (NECA).

Figure 4-6. ELECTRICAL ENERGY SOURCES

Figure 4-6. All electrical energy sources must be accounted for. Courtesy of Ideal Industries, Inc.

Deenergizing Equipment

1910.333(b)(2)(ii)
1910.333(b)(2)(ii)(A)
Safe procedures for deenergizing circuits and equipment shall be determined before circuits or equipment are deenergized.
1910.333(b)(2)(ii)(B)
The circuits and equipment to be worked on shall be disconnected from all electric energy sources. **See Figure 4-6.**

Control circuit devices, such as push buttons, selector switches, and interlocks, may not be used as the sole means for deenergizing circuits or equipment. Interlocks for electric equipment may not be used as a substitute for lockout and tagging procedures.
1910.333(b)(2)(ii)(C)
Stored electric energy which might endanger personnel shall be released. Capacitors shall be discharged and high capacitance elements shall be short-circuited and grounded, if the stored electric energy might endanger personnel.

Note: If the capacitors or associated equipment are handled in meeting this requirement, they shall be treated as energized.
1910.333(b)(2)(ii)(D)
Stored non-electrical energy in devices that could reenergize electric circuit parts shall be blocked or relieved to the extent that the circuit parts could not be accidentally energized by the device. **See Figure 4-7.**

Figure 4-7. STORED ENERGY

Figure 4-7. Stored nonelectrical energy must be blocked or relieved as necessary. Courtesy of Ideal Industries, Inc.

Application of Locks and Tags

1910.333(b)(2)(iii)
1910.333(b)(2)(iii)(A)
A lock and a tag shall be placed on each disconnecting means used to deenergize circuits and equipment on which work is to be performed, except as provided in paragraphs (b)(2)(iii)(C) and (b)(2)(iii)(E) of this section. **See Figure 4-8.**

The lock shall be attached so as to prevent persons from operating the disconnecting means unless they resort to undue force or the use of tools.

1910.333(b)(2)(iii)(B)
Each tag shall contain a statement prohibiting unauthorized operation of the disconnecting means and removal of the tag. **See Figure 4-9.**

1910.333(b)(2)(iii)(C)
If a lock cannot be applied, or if the employer can demonstrate that tagging procedures will provide a level of safety equivalent to that obtained by the use of a lock, a tag may be used without a lock.

1910.333(b)(2)(iii)(D)
A tag used without a lock, as permitted by paragraph (b)(2)(iii)(C) of this section, shall be supplemented by at least one additional safety measure that provides a level of safety equivalent to that obtained by use of a lock. **See Figure 4-10.**

Examples of additional safety measures include the removal of an isolating circuit element, blocking of a controlling switch, or opening of an extra disconnecting device.

1910.333(b)(2)(iii)(E)
A lock may be placed without a tag only under the following conditions:

1910.333(b)(2)(iii)(E)(1)
Only one circuit or piece of equipment is deenergized, and

1910.333(b)(2)(iii)(E)(2)
The lockout period does not extend beyond the work shift, and

1910.333(b)(2)(iii)(E)(3)
Employees exposed to the hazards associated with reenergizing the circuit or equipment are familiar with this procedure.

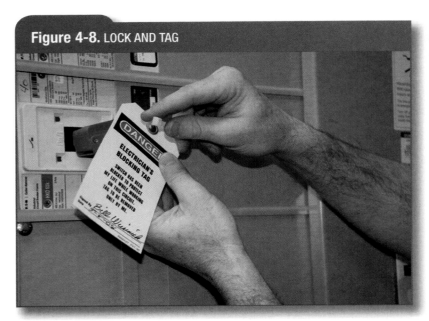

Figure 4-8. LOCK AND TAG

Figure 4-8. *Both a lock and a tag are generally required to be placed on a disconnecting means for circuits and equipment before work begins. Courtesy of National Electrical Contractors Association (NECA).*

Figure 4-9. TAGS

Figure 4-9. *A tag must warn against both unauthorized disconnecting means of operation and unauthorized removal of the tag. Courtesy of Ideal Industries, Inc.*

Verification of Deenergized Condition

1910.333(b)(2)(iv)

The requirements of this paragraph shall be met before any circuits or equipment can be considered and worked as deenergized.

1910.333(b)(2)(iv)(A)

A qualified person shall operate the equipment operating controls or otherwise verify that the equipment cannot be restarted.

1910.333(b)(2)(iv)(B)

A qualified person shall use test equipment to test the circuit elements and electrical parts of equipment to which employees will be exposed and shall verify that the circuit elements and equipment parts are deenergized. **See Figure 4-11.**

The test shall also determine if any energized condition exists as a result of inadvertently induced voltage or unrelated voltage backfeed even though specific parts of the circuit have been deenergized and presumed to be safe. If the circuit to be tested is over 600 volts, nominal, the test equipment shall be checked for proper operation immediately after this test.

Figure 4-10. TAG USED WITHOUT A LOCK

Figure 4-10. Tags are permitted to be used without a lock under limited circumstances, but only where at least one additional safety measure is employed and safety equal to a lock is assured. Courtesy of Ideal Industries, Inc.

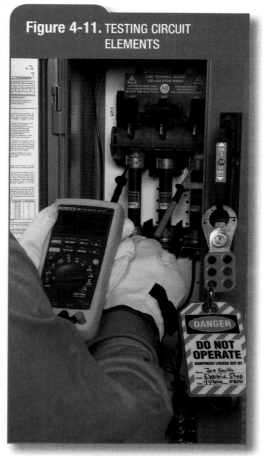

Figure 4-11. TESTING CIRCUIT ELEMENTS

Figure 4-11. A qualified person must test all circuit elements and electrical parts of equipment to verify that they are in a deenergized state.

Reenergizing Equipment

1910.333(b)(2)(v)
These requirements shall be met, in the order given, before circuits or equipment are reenergized, even temporarily.

1910.333(b)(2)(v)(A)
A qualified person shall conduct tests and visual inspections, as necessary, to verify that all tools, electrical jumpers, shorts, grounds, and other such devices have been removed, so that the circuits and equipment can be safely energized.

1910.333(b)(2)(v)(B)
Employees exposed to the hazards associated with reenergizing the circuit or equipment shall be warned to stay clear of circuits and equipment.

1910.333(b)(2)(v)(C)
Each lock and tag shall be removed by the employee who applied it or under his or her direct supervision. **See Figure 4-12**.

However, if this employee is absent from the workplace, then the lock or tag may be removed by a qualified person designated to perform this task provided that:

1910.333(b)(2)(v)(C)(1)
The employer ensures that the employee who applied the lock or tag is not available at the workplace, and

1910.333(b)(2)(v)(C)(2)
The employer ensures that the employee is aware that the lock or tag has been removed before he or she resumes work at that workplace.

1910.333(b)(2)(v)(D)
There shall be a visual determination that all employees are clear of the circuits and equipment.

The Control of Hazardous Energy (Lockout/Tagout)

Many lockout/tagout programs incorporate OSHA 29 CFR 1910.147. This standard is found in Subpart J, General Environmental Controls, of Part 1910, Occupational Safety and Health Standards, and is entitled "The control of hazardous energy (lockout/tagout)."

Recall that Note 2 to1910.333(b)(2) clarifies that lockout and tagging procedures that comply with paragraphs (c) through (f) of 1910.147 will also be deemed to comply with paragraph (b)(2) of this section provided that two criteria are satisfied:

- The procedures address the electrical safety hazards covered by Subpart S.
- The procedures incorporate the requirements of paragraphs 1910.333 (b)(2)(iii)(D) and 1910.333(b)(2) (iv)(B).

Installations in electric power generation facilities that are not an integral part of, or inextricably commingled with, power generation processes or equipment are covered under § 1910.147 and Subpart S of Part 1910.

Figure 4-12. REMOVAL OF LOCKS AND TAGS

Figure 4-12. *Each lock and tag is generally required to be removed by the employee who applied it.* Courtesy of Ideal Industries, Inc.

1910.333(b)(2)(iii)(D)

A tag used without a lock, as permitted by paragraph (b)(2)(iii)(C) of this section, shall be supplemented by at least one additional safety measure that provides a level of safety equivalent to that obtained by use of a lock. Examples of additional safety measures include the removal of an isolating circuit element, blocking of a controlling switch, or opening of an extra disconnecting device.

1910.333(b)(2)(iv)(B)

A qualified person shall use test equipment to test the circuit elements and electrical parts of equipment to which employees will be exposed and shall verify that the circuit elements and equipment parts are deenergized. The test shall also determine if any energized condition exists as a result of inadvertently induced voltage or unrelated voltage backfeed even though specific parts of the circuit have been deenergized and presumed to be safe. If the circuit to be tested is over 600 volts, nominal, the test equipment shall be checked for proper operation immediately after this test.

Section 1910.147 is broken down into six different major headings, each of which is indicated by a different lower-case letter:

a. Scope, application, and purpose
b. Definitions
c. General
d. Application of control
e. Release from lockout or tagout
f. Additional requirements

See 1910.147 for OSHA's requirements for "The control of hazardous energy (lockout/tagout)."

Typical Minimal Lockout Procedures

Appendix A to OSHA 29 CFR 1910.147 serves as a nonmandatory guideline to assist employers and employees in complying with the requirements of 1910.147 and to provide other helpful information. **See Figure 4-13**. Nothing in the appendix adds to or detracts from any of the requirements of 1910.147. Rather, a simple lockout procedure is described to assist employers in developing their own procedures so they meet the requirements of this standard. For more complex systems, more comprehensive procedures may need to be developed, documented, and utilized.

Definitions

OSHA defines the terms *lockout, energy-isolating device, tagout, lockout device,* and *tagout device* in 1910.147(b) and are applicable within 1910.147:

For additional information, visit qr.njatcdb.org Item #1561.

Lockout. The placement of a lockout device on an energy-isolating device, in accordance with an established procedure, ensuring that the energy-isolating device and the equipment being controlled cannot be operated until the lockout device is removed.

Energy-isolating device. A mechanical device that physically prevents the transmission or release of energy, including but not limited to the following: A manually operated electrical circuit breaker; a disconnect switch; a manually operated switch by which the conductors of a circuit can be disconnected from all ungrounded supply conductors, and, in addition, no pole can be operated independently; a line valve; a block; and any similar device used to block or isolate energy. Push buttons, selector switches and other control circuit type devices are not energy isolating devices.

Tagout. The placement of a tagout device on an energy-isolating device, in accordance with an established procedure, to indicate that the energy-isolating device and the equipment being controlled may not be operated until the tagout device is removed.

Lockout device. A device that utilizes a positive means, such as a lock, either key or combination type, to hold an energy-isolating device in the safe position and prevent the energizing of a machine or equipment. Included are blank flanges and bolted slip blinds.

Figure 4-13. EXAMPLE LOCKOUT PROCEDURE

General

The following simple lockout procedure is provided to assist employers in developing their procedures so they meet the requirements of this standard. When the energy isolating devices are not lockable, tagout may be used, provided the employer complies with the provisions of the standard which require additional training and more rigorous periodic inspections. When tagout is used and the energy isolating devices are lockable, the employer must provide full employee protection (see paragraph (c) (3)) and additional training and more rigorous periodic inspections are required. For more complex systems, more comprehensive procedures may need to be developed, documented, and utilized.

Lockout Procedure

Lockout Procedure for

_____

(Name of Company for single procedure or identification of equipment if multiple procedures are used).

Purpose

This procedure establishes the minimum requirements for the lockout of energy isolating devices whenever maintenance or servicing is done on machines or equipment. It shall be used to ensure that the machine or equipment is stopped, isolated from all potentially hazardous energy sources and locked out before employees perform any servicing or maintenance where the unexpected energization or start-up of the machine or equipment or release of stored energy could cause injury.

Compliance With This Program

All employees are required to comply with the restrictions and limitations imposed upon them during the use of lockout. The authorized employees are required to perform the lockout in accordance with this procedure. All employees, upon observing a machine or piece of equipment which is locked out to perform servicing or maintenance shall not attempt to start, energize, or use that machine or equipment.

_____

(Type of compliance enforcement to be taken for violation of the above.)

Sequence of Lockout

(1) Notify all affected employees that servicing or maintenance is required on a machine or equipment and that the machine or equipment must be shut down and locked out to perform the servicing or maintenance.

_____

(Name(s)/Job Title(s) of affected employees and how to notify.)

(2) The authorized employee shall refer to the company procedure to identify the type and magnitude of the energy that the machine or equipment utilizes, shall understand the hazards of the energy, and shall know the methods to control the energy.

_____

(Type(s) and magnitude(s) of energy, its hazards and the methods to control the energy.)

(3) If the machine or equipment is operating, shut it down by the normal stopping procedure (depress the stop button, open switch, close valve, etc.).

_____

(Type(s) and location(s) of machine or equipment operating controls.)

(4) De-activate the energy isolating device(s) so that the machine or equipment is isolated from the energy source(s).

_____

(Type(s) and location(s) of energy isolating devices.)

(5) Lock out the energy isolating device(s) with assigned individual lock(s).

(6) Stored or residual energy (such as that in capacitors, springs, elevated machine members, rotating flywheels, hydraulic systems, and air, gas, steam, or water pressure, etc.) must be dissipated or restrained by methods such as grounding, repositioning, blocking, bleeding down, etc.

_____

(Type(s) of stored energy—methods to dissipate or restrain.)

(7) Ensure that the equipment is disconnected from the energy source(s) by first checking that no personnel are exposed, then verify the isolation of the equipment by operating the push button or other normal operating control(s) or by testing to make certain the equipment will not operate.

Caution: Return operating control(s) to neutral or "off" position after verifying the isolation of the equipment.

_____

(Method of verifying the isolation of the equipment.)

(8) The machine or equipment is now locked out.

"Restoring Equipment to Service." When the servicing or maintenance is completed and the machine or equipment is ready to return to normal operating condition, the following steps shall be taken.

(1) Check the machine or equipment and the immediate area around the machine to ensure that nonessential items have been removed and that the machine or equipment components are operationally intact.

(2) Check the work area to ensure that all employees have been safely positioned or removed from the area.

(3) Verify that the controls are in neutral.

(4) Remove the lockout devices and reenergize the machine or equipment. Note: The removal of some forms of blocking may require reenergization of the machine before safe removal.

(5) Notify affected employees that the servicing or maintenance is completed and the machine or equipment is ready for used.

[54 FR 36687, Sept. 1, 1989 as amended at 54 FR 42498, Oct. 17, 1989; 55 FR 38685, Sept. 20, 1990; 61 FR 5507, Feb. 13, 1996]

Figure 4-13. *This example of a simple lockout procedure, found in OSHA 29 CFR 1910.147 Appendix A, describes how employers may create their own procedures.*

Tagout device. A prominent warning device, such as a tag and a means of attachment, which can be securely fastened to an energy isolating device in accordance with established procedure, to indicate that the energy isolating device and the equipment being controlled may not be operated until the tagout device is removed.

Affected employee. An employee whose job requires him/her to operate or use a machine or equipment on which servicing or maintenance is being performed under lockout or tagout, or whose job requires him/her to work in an area in which such servicing or maintenance is being performed.

OSHA 1910.399 defines the terms *deenergized* and *energized* as follows; these apply to 1910 Subpart S:

Deenergized. Free from any electrical connection to a source of potential difference and from electrical charge; not having a potential different from that of the earth.

Energized. Electrically connected to a source of potential difference.

ACHIEVING AN ELECTRICALLY SAFE WORK CONDITION

OSHA's lockout/tagout requirements are federal law. However, in addition to developing lockout/tagout programs and procedures for working on or near exposed deenergized parts that fulfill the requirements of 29 CFR 1910.147, 1910.333(b), and 1926.417, the process of achieving an electrically safe work condition from *NFPA 70E* Article 120 should be considered, including, but not limited to, *NFPA 70E* Section 120.1.

The Process of Achieving an Electrically Safe Work Condition

The provisions of *NFPA 70E* Article 120 generally meet or exceed OSHA's lockout/tagout requirements and should be examined carefully for considerations related to developing a more comprehensive and complete lockout/tagout program. This includes, but is not limited to, the six-step process outlined in Section 120.1 to verify that an electrically safe work condition exists after the provisions contained in *NFPA 70E* Section 120.2 have been taken into account.

An electrically safe work condition is defined in Article 100 in *NFPA 70E*.

An electrically safe work condition is achieved when the procedures outlined in *NFPA 70E* 120.2 are performed and verified by the six steps outlined in Section 120.1. Achieving an electrically safe work condition is one example of how *NFPA 70E* can supplement and enhance OSHA's requirements for lockout and tagging of circuits (1926.417), control of hazardous energy (lockout/tagout; 1910.147), and the lockout and tagging provisions of 1910.333(b)(2), among others.

Note that Article 120 is divided into three sections that provide the requirements for establishing an electrically safe work condition:

- Verification of an Electrically Safe Work Condition
- Deenergized Electrical Equipment That Has Lockout/Tagout Devices Applied
- Temporary Protective Grounding Equipment

Article 120 needs to be referenced for these requirements in its entirety. Also consider *NFPA 70E* Informative Annex G, Sample Lockout/Tagout Procedure. It is provided for informational purposes only and is not a part of the requirements of *NFPA 70E*. This annex offers a sample procedure to assist employers in developing a procedure that meets the requirements of *NFPA 70E* 120.2.

Verifying an Electrically Safe Work Condition

An electrically safe work condition is achieved when the procedures outlined in *NFPA 70E* Section 120.2 are performed and verified by following the six steps outlined in Section 120.1.

Summary

OSHA's steps for lockout/tagout and detailed requirements for working on or near exposed deenergized parts must be followed. Training, planning, and preparation are critical to avoid incidents and injury.

The Occupational Safety and Health Act of 1970 was enacted "to assure safe and healthful working conditions for working men and women." The provisions of OSHA 29 CFR 1926.417 provide performance requirements for lockout and tagging of circuits within the scope of Subpart K. OSHA CFR 1910.333(b)(2) and OSHA 29 CFR 1910.147 cover minimum steps for lockout/tagout, as well as procedures to follow when lockout or tagout devices are removed and energy is restored within the scope of their respective subparts.

Finally, consider *NFPA 70E* Article 120 to be a supplement to OSHA's requirements and procedures. Following the procedures in *NFPA 70E* Section 120.1 for verification that an electrically safe work condition has been achieved will allow for a more comprehensive lockout/tagout program.

Review Questions

1. OSHA defines a __?__ as a device that utilizes a positive means such as a lock, either key or combination type, to hold an energy isolating device in the safe position and prevent the energizing of a machine or equipment.
   a. lockout device
   b. lockout/tagout device
   c. tagout device
   d. tie wrap

2. OSHA defines a __?__ as a prominent warning device, such as a tag and a means of attachment, which can be securely fastened to an energy isolating device in accordance with established procedure, to indicate that the energy isolating device and the equipment being controlled may not be operated until the device is removed.
   a. labeling device
   b. lockout device
   c. lockout/tagout device
   d. tagout device

3. Per OSHA, each __?__ shall furnish to each employee employment and a place of employment which are free from recognized hazards that are causing or are likely to cause death or serious physical harm to employees.
   a. business manager
   b. employer
   c. trade
   d. training director

4. Each __?__ shall comply with occupational safety and health standards and all rules, regulations, and orders issued pursuant to the OSHA Act which are applicable to his own actions and conduct.
   a. craft
   b. employee
   c. employer
   d. trade

5. Lockout and tagging of circuits, control of hazardous energy, and achieving an electrically safe work condition are among the safe __?__ intended to assure worker safety and health.
   a. habits
   b. happenings
   c. ideas
   d. work practices

6. Per OSHA Part 1926, Subpart C, the employer shall instruct each employee in the recognition and avoidance of __?__ conditions and the regulations applicable to his work environment to control or eliminate any hazards or other exposure to illness or injury.
   a. abnormal
   b. normal
   c. safe
   d. unsafe

7. OSHA Part 1910 states that the lock shall be attached so as to __?__ operation of the disconnecting means unless one would resort to undue force or the use of tools.
   a. allow
   b. discourage
   c. encourage
   d. prevent

8. OSHA Part 1910 states that if a lock __?__ be applied, or if the employer can demonstrate that tagging procedures will provide a level of safety equivalent to that obtained by the use of a lock, a tag may be used without a lock.
   a. can
   b. cannot
   c. should not
   d. will not

9. OSHA Part 1910 states that a tag used without a lock shall be __?__ by at least one additional safety measure that provides a level of safety equivalent to that obtained by use of a lock.
   a. complimented
   b. implemented
   c. prevented
   d. supplemented

Introduction to *NFPA 70E*

Chapter Outline

- *NFPA 70E* History, Introduction, and Application of Safety-Related Work Practices
- General Requirements for Electrical Safety-Related Work Practices
- Establishing an Electrically Safe Work Condition
- Work Involving Electrical Hazards
- Additional Considerations

Chapter Objectives

1. Understand the history, scope, definitions, and organization of *NFPA 70E*.
2. Become familiar with the *NFPA 70E* provisions related to host and contract employer responsibilities and use of electrical equipment.
3. Understand the requirements necessary for creating an electrical safety program, training, and achieving an electrically safe work condition.
4. Become familiar with the topics addressed in Article 130 and the 16 informative annexes in *NFPA 70E*.

Chapter 5

References

1. *NFPA 70E, Standard for Electrical Safety in the Workplace®*, 2015 Edition
2. OSHA 29 CFR Part 1926
3. OSHA 29 CFR Part 1910

Case Study

A 46-year-old electrical project supervisor died when he contacted an energized conductor inside a control panel. The employer was an industrial electrical contracting company that had been in operation for 10 years. It employed 20 workers, including three electrical project supervisors. The company's written safety program, which was administered by the president/CEO and the electrical project supervisors, included disciplinary procedures. The president/CEO served as safety officer on a collateral duty basis, and the supervisors held monthly safety meetings with all crew members.

The victim had worked for the company for five years and three months as an electrical project supervisor and had approximately 27 years of electrical experience. The company and victim had been working at the packaging plant for six months before the incident; the incident was the company's first fatality.

The company had been contracted to install control cabinets, conduit, wiring, and solid-state compressor motor starters for two 400-horsepower air compressors. On the day of the incident, the victim and three coworkers (one electrical worker and two helpers) arrived at the plant at 7 a.m. They were scheduled to install the last starter and to complete the wiring from the compressor motor to the starter in the control panel, and from the starter control panel to the main distribution panel. Once installation was completed, they were to check the operation of the unit.

At approximately 3:15 p.m., the starter had been installed, and all associated wiring had been completed. The victim directed a helper to turn the switch to the "on" position at the main distribution panel, approximately six feet away, to check the starter's operation.

The helper turned the switch to the "on" position, energizing the components inside the starter control panel. The victim pushed the "start" button, and the starter indicator light activated, but the compressor motor did not start. When the compressor motor did not engage, the victim concluded that a problem existed inside the starter control panel. The victim directed the helper to retrieve a voltmeter so that he could check the continuity of the wiring inside the starter control panel. In the interim, the victim opened the starter control panel door without deenergizing the unit and reached inside to trace the wiring and check the integrity of the electrical leads. In doing so, he contacted the 480-volt primary lead for the motor starter with his left hand.

Current passed through the victim's left hand and body and exited through his feet to the ground. The victim yelled, and the helper immediately turned the main distribution switch to the "off" position as the victim collapsed to the floor. Emergency medical services (EMS) was called, and the helper checked the victim and immediately administered cardiopulmonary resuscitation (CPR). EMS personnel arrived in 10 to 15 minutes, continued CPR, and transported the victim to the local hospital, where he was pronounced dead one hour and 20 minutes after the incident occurred.

Source: For details of this case, see FACE Investigation #03CA006. Accessed June 4, 2012.

For additional information, visit qr.njatcdb.org Item #1195.

INTRODUCTION

An important provision in Section 5 of the Occupational Safety and Health Occupational Safety and Health Act of 1970 (OSH Act) requires that workers be provided with a workplace free from recognized hazards. Yet it is not always clear how to provide a hazard-free workplace. OSHA requirements addressing electrical hazards are often written in performance language; that is, the rules define a result without providing details of how to accomplish it. Many people consider the requirements defined in *NFPA 70E* as a means to comply with the OSHA requirements related to the hazards of work involving electrical hazards.

This chapter provides an overview of the *NFPA 70E* standard, with the primary focus on the provisions of Articles 90, 100, 110, and 120. These provisions include, but are not limited to, scope, definitions, host and contract employer responsibilities, training requirements, electrical safety program considerations, use of electrical equipment, and establishing an electrically safe work condition.

NFPA 70E HISTORY, INTRODUCTION, AND APPLICATION OF SAFETY-RELATED WORK PRACTICES

NFPA 70E, Standard for Electrical Safety in the Workplace, consists of three chapters and 16 informative annexes, as well as the Foreword to *NFPA 70E* and Article 90, Introduction. This chapter will focus on Chapter 1, which is divided into five articles: 100, 105, 110, 120, and 130.

Article 100 provides definitions essential to the application of *NFPA 70E*. Article 105 addresses the *application of* safety-related work practices. Article 110 contains the general requirements for electrical safety-related work practices. Article 120 provides the requirements for establishing an electrically safe work condition. Article 130 contains the provisions related to work involving electrical hazards.

History and Evolution of *NFPA 70E*

The Foreword to *NFPA 70E* offers a look at the history and evolution of this NFPA standard. The appointment of the *NFPA 70E* Committee was announced on January 7, 1976. This committee was formed to assist OSHA in preparing electrical safety standards that would serve its requirements and that could be expeditiously promulgated through the provisions of Section 6(b) of the Occupational Safety and Health Act. A primary concern was OSHA's need for electrical regulations that addressed electrical safety-related work practices and maintenance of the electrical system considered critical to safety for employers and employees in their workplaces.

The committee found it feasible to develop a standard for electrical installations that would be compatible with OSHA requirements for safety of the employee in locations covered by the *National Electrical Code*® (*NEC*). The new standard was named *NFPA 70E, Standard for Electrical Safety Requirements for Employee Workplaces*. The first edition was published in 1979.

The fifth edition, published in 1995, included the concepts of "limits of approach" and the establishment of an "arc." In 2000, the newly published sixth edition continued to focus on establishment of flash protection boundaries, the use of personal protective equipment, and charts to assist the user in applying appropriate protective clothing and personal protective equipment for common tasks.

The seventh edition, published in 2004, reflected a name change of the document to *NFPA 70E, Standard for Electrical Safety in the Workplace*, as well as the addition of the energized electrical work permit and related requirements.

The 2015 edition of *NFPA 70E*, Standard for Electrical Safety in the Workplace, has an effective date of July 29, 2014, and supersedes all previous editions.

NFPA 70E Introductory Information

Article 90 contains the introductory information for *NFPA 70E*, Standard for Electrical Safety in the Workplace. It spells out the purpose, scope, arrangement, and organization of the *70E* standard. Also covered here are the requirements related to formal interpretations as well as those related to mandatory rules, permissive rules, and explanatory material. Refer to Article 90 for these requirements in their entirety.

Purpose and Scope

The purpose of *NFPA 70E* is the first requirement in Article 90, Introduction. It is to provide a practical safe working area for employees relative to the hazards arising from the use of electricity. The scope in Article 90 outlines what is covered and not covered by this standard.

Organization and Arrangement

The *NFPA 70E* standard is divided into Article 90, Introduction, 3 chapters, and 16 informative annexes.

Application of the three chapters of the *NFPA 70E* standard are described in 90.3, Standard Arrangement. Annexes are not part of the requirements of this standard but are included for informational purposes only.

Rules, Explanatory Material, and Formal Interpretations

Section 90.5 spells out the provisions in *NFPA 70E* related to mandatory rules, permissive rules, and explanatory material. Mandatory rules of the *NFPA 70E* standard identify actions that are specifically required or prohibited. Permissive rules, in contrast, identify actions that are allowed but not required.

Explanatory material is included in the form of informational notes, which are not enforceable as requirements. Brackets containing section references to another NFPA document are for informational purposes only and are provided as a guide to indicate the source of the extracted text.

Formal interpretation procedures have been established and are found in the NFPA Regulations Governing Committee Projects.

NFPA 70E Definitions

The scope of Article 100 indicates that only those definitions deemed essential to the proper application of *NFPA 70E* are provided there. The definitions apply wherever the terms are used throughout *NFPA 70E*.

Refer to Article 100 for these definitions. Become familiar with each defined term, including any informational notes, and apply it within it meaning each time it is used throughout *NFPA 70E*. While there are a number of definitions that are seldom if ever applied, there are a number of definitions that are used frequently and their meaning is essential in understanding and properly applying *NFPA 70E*. These include arc flash hazard, arc rating, balaclava, arc flash boundary (boundary, arc flash), limited approach boundary (boundary, limited approach), restricted approach boundary (boundary, restricted approach), de-energized, electrical hazard, electrically safe work condition, energized, incident energy analysis, qualified person, risk assessment, shock hazard, unqualified person, nominal voltage, and working on.

Application of Safety-Related Work Practices

The title of Chapter 1 of *NFPA 70E* is Safety-Related Work Practices. Not only are definitions contained in this chapter but so too are the rules for the application of the Chapter 1 safety-related work practices. These provisions are housed in Article 105, Application of Safety-Related Work Practices. Refer to Article 105 for the application provisions related to the scope, purpose, responsibility (employer and employee), and organization in their entirety.

GENERAL REQUIREMENTS FOR ELECTRICAL SAFETY-RELATED WORK PRACTICES

NFPA 70E Chapter 1 also contains general requirements for electrical safety-related work practices. These general requirements are located in Article 110. The following topics are addressed in this article:

- Electrical safety program
- Training requirements
- Emergency response training
- Host and contract employer's responsibilities
- Use of electrical equipment

A brief overview of the five topics in the Article 110 general requirements for electrical safety-related work practices follow. Article 110 needs to be referenced for these requirements in their entirety.

Electrical Safety Program

Many who open the pages of *NFPA 70E* expect to find a comprehensive, ready-to-go electrical safety program provided. What *NFPA 70E* actually provides is not a complete electrical safety program, but rather a number of minimum considerations that must be evaluated and integrated into the framework of an existing electrical safety program to make it more comprehensive. Specific requirements are provided in 110.1 as to what must be included in an *NFPA 70E* compliant electrical safety program. These are divided into nine major categories. Non-mandatory information is also provided in *NFPA 70E* Informative Annex E, Electrical Safety Program.

General, Maintenance, and Awareness and Self-Discipline Requirements. The first of nine considerations for a safety program are the general requirements. Here it is required, in part, that the employer must "implement and document an electrical safety program that directs activity appropriate to the risk associated with electrical hazards." There are also four informational notes offering insight into these requirements including a reference to ANSI/AIHA Z10.

The second of nine considerations are those requiring that the electrical safety program "include elements that consider the condition of maintenance of electrical equipment and systems." This is just one of a number of locations in *NFPA 70E* where the condition of maintenance must be considered. Also see 130.5(3), 130.5 Informational Note No. 1, Table 130.7(C)(15)(A)(a), and 205.3, for example.

The third of nine factors that *NFPA 70E* requires to be included in the electrical safety program is awareness and self-discipline. These requirements are located in 110.1(C).

Electrical Safety Program Principles, Controls, and Procedures. The fourth, fifth, and sixth of the nine factors that *NFPA 70E* requires to be included in the electrical safety program are program principles, controls, and procedures. The electrical safety program must identify "the principles upon which it is based, the controls by which it is measured and monitored, and the procedures that are to be utilized before work is started by employees exposed to an electrical hazard. No specific requirements indicate what those program principles, controls, and procedures must be. While reference is made to the information in Informative Annex E.1, E.2, and E.3 through informational notes, these notes and annex information are non-mandatory. Informative Annexes E.1, E.2, and E.3 include examples of considerations as part of the required principles, controls, and procedures.

Risk Assessment Procedure. The seventh element required by *NFPA 70E* to be included in an electrical safety program is a risk assessment procedure. In part, the procedure requires identification of hazards, assessment of risks, and implementation of risk control. The third in the list, risk control implementation, identifies a six method hierarchy from ANSI/AIHA Z10. Although many might turn to PPE as their first resort, note that PPE is the last of those six methods in this hierarchy.

Additional guidance is offered in other informational notes to these requirements. The second of three notes discusses ". . . identifying when a second person might be required and the training and equipment that second person should have." The third of three informational notes here in 110.1(G) mentions Informative Annex F as an example of a risk assessment procedure.

Job Briefing. The eighth of nine topics that *NFPA 70E* requires to be addressed in an electrical safety program is the job briefing. The job briefing requirements mandate that job briefings be conducted by the employee in charge before commencement of work and include everyone who will be involved. Additional job briefings are required if changes that might affect the safety of employees occur during the course of the work.

The requirements in 110.1(H) also mention examples of the subjects that must be covered during a job briefing. One of these subjects is the information on the energized electrical work permit, if required.

The informational note alerts that Figure I.1 in Informative Annex I offers an example of a job briefing form and checklist. Keep in mind that Informative Annex I is non-mandatory in both format and content.

Electrical Safety Auditing. The last of the nine topics that *NFPA 70E* requires to be included in an electrical safety program is documented program auditing. Per 110.1(I)(1), the electrical safety program must be audited to verify that the principles and procedures of the electrical safety program are in compliance with audits performed at a frequency of not more than three years. Field work audits are addressed in (2). Among the provisions here is the requirement that field audits "be performed at intervals not to exceed 1 year." In accordance with (3), audits required by 110.1(I) must be documented.

Training Requirements

Training requirements are located in 110.2. These requirements are divided into five major headings.

- Safety training
- Type of training
- Emergency response training
- Employee training
- Training documentation

Safety Training and Type of Training. The first of five headings for training requirements address "safety training." The training mentioned in 110.2(A) covers, in part, employee training to "understand the specific hazards associated with electrical energy," training "in safety-related and procedural requirements, as necessary, to provide protection against electrical hazards . . .", and training "to identify and understand the relationship between electrical hazards and possible injury."

The subject of the second of five training requirement headings is "type of training." The provisions of 110.2(B) recognize that the training required by 110.2 can be classroom training, on-the-job training, or a combination of these. Regardless, the "type and extent of the training provided shall be determined by the risk to the employee."

Emergency Response Training. The third of five headings for training requirements address what is categorized as "emergency response training." This third category of training is subdivided into four sub-categories:

- Contact release
- First aid, emergency response, and resuscitation
- Training verification
- Documentation

The various training requirements under these four sub-categories are located in 110.2(C) in *NFPA 70E*. The training requirements related to contact release, including training "in methods of safe release of victims from contact . . ." are addressed in (C)(1). Those for first aid, emergency response, and resuscitation are in (C)(2), those requiring the employer to verify that training is current at least annually is in (C)(3), and the requirement that all emergency response training be documented is in (C)(4). These, like all 70E requirements, need to be referenced and applied in their entirety.

Employee Training and Training Documentation. The fourth of five headings for training requirements address "employee training" and are found in 110.2(D) in *NFPA 70E*. The requirements for "qualified person" employee training are located in (D)(1). These training requirements are quite extensive and, like all NFPA 70E requirements, need to be reviewed and applied in their entirety.

The requirements spelling out the training that unqualified persons are required to have are in (D)(2). While much less detailed than those for qualified person employee training, the training requirements for unqualified persons are no less important. Unqualified persons must "be trained in, and be familiar with, any electrical safety-related practices necessary for their safety."

The fifth and last of the five headings under training requirements addresses training documentation. These requirements in 110.2(E) require, in part, that the training provided to both qualified persons and unqualified persons be documented when the employee demonstrates proficiency, be maintained for the duration of their employment, and that the documentation include the content of the training, each employee's name, and dates of training.

Host and Contract Employer Responsibilities

The Article 110 general requirements that address both host and contractor employers' responsibilities are located in 110.3(A) and (B). Meeting documentation requirements are located in (C). A documented meeting between the host employer and the contract employer may be required as part of these joint responsibilities.

Among the host employer responsibilities is a requirement to provide the contract employer with the information about the installation necessary to make the assessments required by Chapter 1 of *NFPA 70E*. Among the contract employer responsibilities is a requirement that the contract employer ensure that each employee follows the work practices required by NFPA 70E in addition to any safety-related work rules of the host employer.

Consult 110.3 to understand and apply all of the host employer, contract employer, and related documentation requirements, as applicable.

Use of Electrical Equipment

Familiarity with the *NFPA 70E* requirements related to the use of electrical equipment is also important because tasks where these requirements might apply are performed with some fre-

quency. 70E 110.4 addresses five major topics:

- Test instruments and equipment
- Portable electric equipment
- Ground-fault circuit interrupter (GFCI) protection
- GFCI protection devices
- Overcurrent protection modification

Test Instruments and Equipment. The first of these five categories is covered in 110.4(A). This category covering test instruments and equipment is further divided into five subcategories: (1) testing, (2) rating, (3) design, (4) visual inspection and repair, and (5) operation verification. Informational note guidance to 110.4(A)(2) references ANSI/ISA-61010-1 (82.02.01)/UL 61010-1, Safety Requirements for Electrical Equipment for Measurement, Control, and Laboratory Use—Part 1: General Requirements, for rating and design requirements for voltage measurement and test instruments intended for use on electrical systems of 1,000 volts or less. **See Figure 5-1.** It is important to review and apply all of these requirements. The use of test instruments and equipment is a fundamental and important task that an electrical worker is often expected to perform.

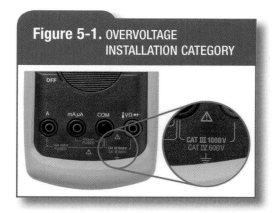

Figure 5-1. *The Overvoltage Installation Category is one of the ratings to be considered for test equipment.* Courtesy of Fluke Corporation.

Compare what is required by 110.4(A)(1),(2) and (4) with what OSHA requires in 1910.334(c) for test instruments and equipment:

For additional information, visit qr.njatcdb.org Item #1562.

> 1910.334(c) Test instruments and equipment.
>
> 1910.334(c)(1) Use. Only qualified persons may perform testing work on electric circuits or equipment.
>
> 1910.334(c)(2) Visual inspection. Test instruments and equipment and all associated test leads, cables, power cords, probes, and connectors shall be visually inspected for external defects and damage before the equipment is used. If there is a defect or evidence of damage that might expose an employee to injury, the defective or damaged item shall be removed from service, and no employee may use it until repairs and tests necessary to render the equipment safe have been made.
>
> 1910.334(c)(3) Rating of equipment. Test instruments and equipment and their accessories shall be rated for the circuits and equipment to which they will be connected and shall be designed for the environment in which they will be used.

OSHA Tip

Qualified person. One who has received training in and has demonstrated skills and knowledge in the construction and operation of electric equipment and installations and the hazards involved.

Note 1 to the definition of "qualified person:" Whether an employee is considered to be a "qualified person" will depend upon various circumstances in the workplace. For example, it is possible and, in fact, likely for an individual to be considered "qualified" with regard to certain equipment in the workplace, but "unqualified" as to other equipment. (See 1910.332(b)(3) for training requirements that specifically apply to qualified persons.)

Note 2 to the definition of "qualified person:" An employee who is undergoing on-the-job training and who, in the course of such training, has demonstrated an ability to perform duties safely at his or her level of training and who is under the direct supervision of a qualified person is considered to be a qualified person for the performance of those duties.

[1910.399, definitions applicable to OSHA 29 CFR Part 1910 Subpart S]

Portable Electric Equipment. The second of five categories in 110.4(B) addresses the use of cord- and plug-connected equipment, including cord sets (extension cords) and covers four main topics:

- Handling and storage
- Grounding-type equipment
- Visual inspection and repair of portable cord- and plug-connected equipment and flexible cord sets
- Conductive work locations
- Connecting attachment plugs
- Manufacturer's instructions

Be sure to review and apply these requirements from 110.4(B). Consider how they differ from or can supplement any applicable OSHA requirements and company safety rules.

GFCI Protection, GFCI Protection Devices, and Overcurrent Protection Modification. The third, fourth and fifth of five categories in 110.4, Use of Electrical Equipment, address GFCI protection, GFCI devices, and overcurrent protection modification.

The rules in 110.4(C) covering GFCI protection are divided into three subcategories: (1) general, (2) maintenance and construction, and (3) outdoors. Be sure to review and apply these requirements from 110.4(C)(1), (2) and (3). Consider how they differ from or can supplement any applicable NEC or OSHA requirements and company safety rules.

The provisions in 110.4(D) require GFCI protection devices to be tested in accordance with the manufacturer's instructions.

NFPA 70E requirements related to use of electrical equipment conclude by addressing modification of overcurrent protection in 110.4(E). Overcurrent protection of circuits and conductors is not permitted to be modified, even on a temporary basis, beyond that permitted by applicable portions of electrical codes and standards dealing with overcurrent protection.

ESTABLISHING AN ELECTRICALLY SAFE WORK CONDITION

An electrically safe work condition is defined in Article 100 of *NFPA 70E*. The requirements for establishing an electrically safe work condition are located in Article 120. This article is divided into three headings:

- Verification of an electrically safe work condition
- De-energized electrical equipment that has lockout/tagout devices applied
- Temporary protective grounding equipment

Verification of an Electrically Safe Work Condition

Article 120 begins by describing a six-step process to verify that an electrically safe work condition has been achieved. The provisions of 120.1 indicate that "an electrically safe work condition shall be achieved when performed in accordance with the procedures of 120.2 and verified by . . ." the six step process in 120.1. Note that an electrically safe work condition has not been achieved just by following the six step process in 120.1. This six step process is intended as verification that an electrically safe work condition has been achieved when the procedures of 120.2 have been performed.

De-energized Electrical Equipment That Has Lockout/Tagout Devices Applied

The requirements of 120.2 are comprehensive and span nearly four pages in *NFPA 70E*. The requirements of 120.2, De-energized Electrical Equipment That Has Lockout/Tagout Devices Applied, require, in part, that each employer "shall identify, document, and implement lockout/tagout procedures conforming to Article 120 . . ." The nearly four pages of 120.2 requirements are divided into six major headings:

- General
- Principles of lockout/tagout execution
- Responsibility
- Hazardous electrical energy control procedure
- Equipment
- Procedures

General. The general requirements are located in 120.2(A). Among other things, these general provisions describe the dis-

tinction between energized, de-energized, disconnected, and electrically safe, for example.

Principles of Lockout/Tagout Execution. Principles of lockout/tagout execution is the second of six topics addressed in 120.2. These principles are covered in 120.2(B)—which is further divided into nine categories:

1. Employee involvement
2. Training
3. Retraining
4. Training documentation
5. Plan
6. Control of energy
7. Identification
8. Voltage
9. Coordination

Review these principles of lockout/tagout execution thoroughly. Compare them to those required in OSHA-based lockout, tagging, and control of hazardous energy requirements.

Responsibility. Responsibility is the third of six topics addressed in 120.2. These responsibilities are covered in 120.2(C) and are divided into three subcategories: (1) procedures, (2) form of control, and (3) audit procedures. An informational note to 120.2(C)(2) points to Informative Annex G for an example of a lockout/tagout procedure. Consult all of the provisions of 120.2(C) in their entirety for a complete understanding of the responsibility requirements as they apply within the provisions of 120.2, de-energized electrical equipment that has lockout/tagout devices applied.

Hazardous Electrical Energy Control Procedure. This is the fourth of six topics addressed in 120.2. The requirements here in 120.2(D), Hazardous Electrical Energy Control Procedure, are subdivided into three subcategories: (1) simple lockout procedure, (2) complex lockout/tagout procedure, and (3) coordination. These requirements describe these procedures, when and how they apply, and how they are to be coordinated, for example. Refer to these procedures in their entirety for a full and complete understanding of their content and application.

Figure 5-2. ROLE OF EQUIPMENT

Figure 5-2. Equipment, such as locks and tags, play an important role in achieving an electrically safe work condition. *Courtesy of Ideal Industries, Inc.*

For additional information, visit qr.njatcdb.org Item #1561.

For additional information, visit qr.njatcdb.org Item #1563.

Equipment. Equipment is the fifth of six topics addressed in 120.2. **See Figure 5-2.** The equipment provisions of 120.2(E) are divided into six categories:

1. Lock application
2. Lockout/tagout device
3. Lockout device
4. Tagout device
5. Electrical circuit interlocks
6. Control devices

These six subcategories need to be reviewed thoroughly to understand these comprehensive rules in 120.2(E), Equipment, and their application within 120.2, De-energized Electrical Equipment That Has Lockout/Tagout Devices Applied.

Procedures. Procedures is the last of the six topics within 120.2. Per 120.2(F), the employer must maintain a copy of these procedures and make them available to employees. The requirements for procedures are divided into those for (1) planning and (2) elements of control.

The planning requirements in (F)(1) are further divided into five categories:

- Locating sources
- Exposed persons
- Person in charge
- Simple lockout/tagout
- Complex lockout/tagout

The elements of control provisions in (F)(2) are further divided into fourteen categories:

- De-energizing equipment (shutdown)
- Stored energy
- Disconnecting means
- Responsibility
- Verification
- Testing
- Grounding
- Shift change
- Coordination
- Accountability
- Lockout/tagout application
- Removal of lockout/tagout devices
- Release for return to service
- Temporary release for testing/positioning

Note that the requirements for planning in 120.2(F)(1) not only require the procedure for planning to include everything addressed in (F)(1) but to also include everything in 120.2(F)(2), Elements of Control, as well.

Note too that many of these provisions are similar to what OSHA sets forth in 1910.147 and 1910.333(b)(2). Be sure to review all of the requirements in 120.2(F) in their entirety. Compare these procedures with what OSHA requires in its related requirements.

Temporary Protective Grounding Equipment

Article 120 concludes with requirements for temporary protective grounding equipment. **See Figure 5-3.** These rules apply within Article 120, Establishing An Electrically Safe Work Condition. The requirements of 120.3 are divided into four subcategories:

1. Placement
2. Capacity
3. Equipment approval
4. Impedance

These temporary protective grounding equipment requirements are an important part of Article 120. See 120.1(6) and 120.2(F)(2)(g), for example. However, all provisions required to establish an electrically safe work condition need to be reviewed and applied in their entirety.

In addition, although not specifically mentioned in 130.7(D), Other Protective Equipment, note that temporary protective grounds are among the subjects addressed in Table 130.7(F), Standards on Other Protective Equipment.

WORK INVOLVING ELECTRICAL HAZARDS

Article 130 contains the provisions related to work involving electrical hazards. In fact, "work involving electrical hazards" is the title of Article 130. Review the definition of electrical hazard in Article 100 of *NFPA 70E*. Per this definition it is clear that Article 130 covers work involving four hazards caused by either contact or equipment failure.

Article 130 is among the most widely used articles in *NFPA 70E*. An entire chapter in this publication will be dedicated to examining Article 130 in some detail. A high level overview of Article 130 will be provided here as an introduction to what is address in that article.

Note that Article 130 is divided into the following ten categories, which provide the framework for the requirements for work involving electrical hazards:

- General
- Electrically Safe Working Conditions
- Working While Exposed to Electrical Hazards
- Approach Boundaries to Energized Electrical Conductors or Circuit Parts for Shock Protection
- Arc Flash Risk Assessment
- Other Precautions for Personnel Activities
- Personal and Other Protective Equipment
- Work Within the Limited Approach Boundary or Arc Flash Boundary of Overhead Lines
- Underground Electrical Lines and Equipment
- Cutting or Drilling

The next chapter in this publication provides an expanded look at these and other topics. *NFPA 70E* needs to be referenced for the Article 130 provisions in their entirety.

Figure 5-3. GROUND-CONNECTING DEVICES

Figure 5-3. *Apply ground-connecting devices rated for the available fault duty as warranted.* Photo courtesy of National Electrical Contractors Association (NECA).

ADDITIONAL CONSIDERATIONS

It is important to remember that Section 90.3 provides the following arrangement of the standard:

- Chapter 1 applies generally for safety-related work practices.
- Chapter 2 applies to safety-related maintenance requirements for electrical equipment and installations in workplaces.
- Chapter 3 supplements or modifies Chapter 1 with safety requirements for special equipment.

As an overview, the following is a list of the three chapters, their articles, and the topics addressed in each:

Chapter 1: Safety-Related Work Practices
    Article 100: Definitions
    Article 105: Application of Safety-Related Work Practices
    Article 110: General Requirements for Electrical Safety-Related Work Practices
    Article 120: Establishing an Electrically Safe Work Condition
    Article 130: Work Involving Electrical Hazards
Chapter 2: Safety-Related Maintenance Requirements
    Article 200: Introduction
    Article 205: General Maintenance Requirements

While *NFPA 70E* contains 16 informative annexes, these are not part of the requirements of the standard but rather are included for informational purposes only.

The following list identifies the 16 informative annexes and their topics:

Summary

This chapter provides an overview of the *NFPA 70E* standard, with a primary focus on the provisions of Articles 90, 100, 110, and 120. These provisions include, but are not limited to, host and contract employer responsibilities, training requirements, electrical safety program requirements, use of electrical equipment, and establishing an electrically safe work condition. Understanding the definitions from Article 100 and the requirements of Articles 110, 120, and 130 is essential in any effort to comply with these requirements. Recognize that Chapters 2 and 3 of *NFPA 70E* also contain numerous additional electrical safety requirements. The entire *NFPA 70E* standard, including the information within the sixteen informative annexes, and all applicable OSHA requirements need to be reviewed and considered. An effective electrical safety program, training, and work practices will adopt and implement appropriate requirements from all sources of information.

Summary cont.

An electrical safety program and effective training provide a foundation for a workplace free from recognized hazards. Achieving an electrically safe work condition is the primary safety-related work practice and must be honored unless the employer demonstrates infeasibility or a greater hazard. Additional safety-related work practices must be determined through both a shock risk assessment and an arc flash risk assessment before any person is exposed to electrical hazards. Energized work may include tasks such as verifying the absence of voltage to even place equipment in an electrically safe work condition.

The importance of establishing an electrical safety program and appropriate training cannot be overstated. An electrical safety program is essential as part of an effort to train and protect workers from electrical hazards and a fundamental component of providing an effective overall safety and health program for worker safety. It is important to review and understand the nine elements that *NFPA 70E* requires to be included in an electrical safety program as minimum considerations for making improvements to a new or existing electrical safety program.

Review Questions

1. *NFPA 70E* requires the employer to implement and document a(n) __?__ program.
   a. cost benefit analysis
   b. electrical safety
   c. loss prevention
   d. research and development

2. Article __?__ covers safety-related maintenance requirements for fuses and circuit breakers.
   a. 205
   b. 225
   c. 340
   d. 350

3. Article __?__ outlines what is covered and not covered by *NFPA 70E*.
   a. 90
   b. 100
   c. 110
   d. 120

4. *NFPA 70E* requires that emergency response training be documented.
   a. True
   b. False

5. The scope of Article 100 indicates that each term or phrase used within *NFPA 70E* is defined in Article 100 when its meaning is not likely to be well understood.
   a. True
   b. False

6. An unqualified person must be trained in any safety-related work practices necessary for their safety and this training must be documented and on file for the duration of their employment.
   a. True
   b. False

7. The host employer is required to provide the contract employer with the information about the installation necessary to make the assessments required by Chapter 1 of *NFPA 70E*.
   a. True
   b. False

8. The title of ANSI/ISA 61010-1 is "Safety Requirements for Electrical Equipment for Measurement, Control, and Laboratory Use, Part 1: General Requirements, for rating and design requirements for voltage measurement and test instruments intended for use on electrical systems 1000 volts and below."
   a. True
   b. False

9. Article __?__ contains the provisions related to work involving electrical hazards.
   a. 100
   b. 130
   c. 200
   d. 330

10. The title of Informative Annex J is __?__.
    a. Electrical Safety Program
    b. Energized Electrical Work Permit
    c. General Categories of Electrical Hazards
    d. Guidance on Selection of Protective Clothing and Other Personal Protective Equipment (PPE)

Justification, Assessment, and Implementation of Energized Work

Chapter Outline

- Application, Justification, and Documentation
- Working While Exposed to Electrical Hazards
- Approach Boundaries to Energized Electrical Conductors or Circuit Parts for Shock Protection
- Arc Flash Risk Assessment
- Other Precautions for Personnel Activities
- Personal and Other Protective Equipment
- Overhead Lines
- Underground Electrical Lines and Equipment
- Cutting or Drilling

Chapter Objectives

1. Recognize when energized work is permitted and the elements of an energized electrical work permit.
2. Understand the requirements related to approach boundaries for shock protection and those for an arc flash risk assessment.
3. Understand the requirements for personal and other protective equipment.
4. Know the requirements related to precautions for personnel activities, alerting techniques, and work within the limited approach or arc flash boundary of overhead lines, underground electrical lines and equipment, and cutting and drilling.

Chapter 6

Reference

1. *NFPA 70E*®, 2015 Edition

Case Study

A 60-year-old electrical worker was electrocuted when he contacted an energized 277-volt circuit. The incident occurred in an office building that was being renovated.

The employer had no electrical training or safety program, explaining that the workers were hired from the union hall and were certified by the union as trained Journeymen or apprentices. As part of the contract with the site owner, the employer was required to follow all of the site owner's safety rules, including the enforcement of a lockout/tagout procedure.

Three employees of the electrical contractor were at the site on the morning of the incident: the foreman, a Journeyman electrician (the victim), and an apprentice. At 7:15 a.m., the foreman gave the victim a set of blueprints and explained the job to him. He was told to install new two-by-two-foot fluorescent lighting fixtures in two offices that were being expanded. He was instructed to install a fixture tail to one fixture and then to connect it to the existing lights. The Journeyman electrician was also told to reconnect the power on a second bank of lights in an adjacent area that had been disconnected a week earlier. The foreman did not instruct the victim to deenergize the circuit breakers.

The Journeyman and the apprentice went to the site. At some point, the victim asked the apprentice, "What's up there?" The apprentice replied, "There are a hot switch leg and another fixture tail to be tied into the new fixtures," to which the victim said, "Okay." The switch leg was an energized 277-volt electrical cable that had been previously connected to a wall switch in a partition wall. The fixture tail was an electrical cable from another fixture that had not yet been connected and was deenergized.

At approximately 9:30 a.m., after a coffee break, the victim resumed work on wiring the fixtures. He contacted the 277 volts apparently while stripping the insulation from the switch leg. The electric shock burned the victim's left hand and knocked him from the ladder. The apprentice heard the Journeyman fall and went to his aid. He checked the victim for a pulse and found none. An office employee then started cardiopulmonary resuscitation (CPR). The paramedics arrived soon after, and the victim was transported to the local hospital, where he was pronounced dead at 10:36 a.m.

It is not known why the victim was working on the energized switch leg. The apprentice thought that the victim grabbed the wrong cable and failed to test both the black and white wires. The foreman later stated that there was no reason for the circuits to be energized and that the crew always tested every circuit beforehand. Although the contractor had access to the breaker, no one had deenergized the circuits in that area. The Occupational Safety and Health Administration (OSHA) noted that the second bank of lights adjacent to the accident site was deenergized by taping off the wall switch.

The county medical examiner attributed the cause of death to electrocution.

Source: For details of this case, see New Jersey Case Report 92NJ007. Accessed July 19, 2012.

For additional information, visit qr.njatcdb.org Item #1196.

INTRODUCTION

Article 130 contains the provisions related to work involving electrical hazards. As described earlier, an electrical hazard is defined in *NFPA 70E*.

Therefore, by definition, Article 130 covers work involving four hazards caused by either contact or equipment failure. **See Figure 6-1.**

Article 130 is divided into ten sections that provide the requirements for work involving electrical hazards:
- General
- Electrically Safe Working Conditions
- Working While Exposed to Electrical Hazards
- Approach Boundaries to Energized Electrical Conductors or Circuit Parts for Shock Protection
- Arc Flash Risk Assessment
- Other Precautions for Personnel Activities
- Personal and Other Protective Equipment
- Work Within the Limited Approach Boundary or Arc Flash Boundary of Overhead Lines
- Underground Electrical Lines and Equipment
- Cutting or Drilling

An overview, excerpts, and abbreviated content from the requirements for work involving electrical hazards follow. Article 130 needs to be referenced for a complete understanding of these requirements in their entirety.

APPLICATION, JUSTIFICATION, AND DOCUMENTATION

The first part of Article 130 addresses what might be categorized as application, justification, and documentation through the general requirements in 130.1, the conditional requirement that energized conductors and circuit parts be put into an electrically safe work condition before work is performed in 130.2, justification for energized work in 130.2(A), and the documentation and other requirements of the energized electrical work permit in 130.2(B).

Figure 6-1. ELECTRICAL HAZARDS

Figure 6-1. *An electrical hazard is a dangerous condition that may be caused by contact or equipment failure.* Courtesy of Westex.

General

Article 130 begins with the general requirement in 130.1 stating the two things that Article 130 covers. The first is when an electrically safe work condition must be established. The second thing that Article 130 covers is electrically safety-related work practices when an electrically safe work condition cannot be established. These general requirements in 130.1 also stipulate that all of the requirements of Article 130 apply regardless of whether an incident energy analysis is completed or if the arc flash PPE categories method is used for the selection of arc flash PPE in lieu of an incident energy analysis. See 130.5, 130.5(C), and 130.5(C)(1) and (2).

Review the definition of incident energy analysis from Article 100 to aid in the understanding and application of this rule. Note that it is a component of an arc flash risk assessment. Refer to 130.5 for all of what is required as part of an arc flash risk assessment such as the documentation requirements in (A), the arc flash boundary requirements in (B), the arc flash PPE requirements in (C), and the equipment labeling requirements in (D).

The significance of this "all requirements of this article shall apply" rule may not be immediately apparent, as these referenced requirements are covered somewhat later in Article 130. As an overview, recognize that incident energy analysis or Table 130.7(C)(15)(A)(a), Table 130.7(C)(15)(A)(b), Table 130.7(C)(15)(B), and Table 130.7(C)(16) are the two ways recognized to select arc flash PPE per 130.5(C). Consider, for example, whether the provisions of 130.7(C), such as 130.7(C)(1), 130.7(C)(6), 130.7(C)(9), and 130.7(C)(10), apply when the "Arc Flash PPE Categories Method" is used.

This rule might appear to apply only to an incident energy analysis and the referenced tables. In reality, all of the Article 130 provisions—such as those related to when energized work is permitted, when an energized electrical work permit is required, working while exposed to electrical hazards, approach boundaries for shock protection, arc flash risk assessment, other precautions for personnel activities, personal and other protective equipment, underground and overhead lines, and cutting and drilling—apply no matter whether an incident energy analysis is completed or the referenced tables are used. Other examples of the application of this rule are provided later in this chapter when the requirements for incident energy analysis and the referenced tables are covered in detail. Also keep in mind that the definitions in Article 100, the general requirements for electrical safety related work practices in Article 110, and the rules for establishing an electrically safe work condition in Article 120 apply to Article 130 as well.

Electrically Safe Working Conditions and Energized Work

The second of the ten sections in Article 130, titled Electrically Safe Working Conditions, covers rules for the need for an electrically safe work condition versus energized work and when an energized electrical work permit is required. Section 130.2 specifies when energized electrical conductors and circuit parts must be put into an electrically safe work condition. Also review the exception to this rule.

Energized work is addressed in 130.2(A). There are four subcategories that make up this requirement:

1. Additional hazards or increased risk
2. Infeasibility
3. Less than 50 volts
4. Normal operation

Carefully review these rules outlining energized work justification as well as the informational notes that provide additional insight into the energized work requirements of 130.2(A). **See Figure 6-2.**

Energized Electrical Work Permit

The provisions for an energized electrical work permit are set forth in 130.2(B). Only after satisfying one of the four conditions for energized work in 130.2(A) may energized work take place. The provisions in 130.2(B)(1) then provide the two conditions when an energized electrical work permit is required. These two conditions are set forth in 130.2(B)(1)(1) and (B)(1)(2). As the requirements of 130.2(B) are reviewed in detail, be sure to also note the conditions where a permit would be required even when the conductors or circuit parts are "not exposed" as provided for in 130.2(B)(1)(2).

Now that it has been determined when an energized electrical permit is required per 130.2(B), explore the eight headings of the minimum considerations that must be included in an energized electrical work permit.

The first required element asks that what is being worked on and its location be documented. The second element asks that the justification for energized work from 130.2(A) be documented. The third element asks for a description of the safe work practices that will be used by qualified persons to account for any hazards identified in the shock risk assessment and arc flash risk assessment required by 130.3, for example.

Note that there are a number of OSHA and *NFPA 70E* rules that require hazard documentation. Recall OSHA's provisions of 1910.132(d)(1) and (d)(2):

1910.132(d)(1)

The employer shall assess the workplace to determine if hazards are present, or are likely to be present, which necessitate the use of personal protective equipment (PPE).

1910.132(d)(2)

The employer shall verify that the required workplace hazard assessment has been performed through a written certification that identifies the workplace evaluated; the person certifying that the evaluation has been performed; the date(s) of the hazard assessment; and, which identifies the document as a certification of hazard assessment.

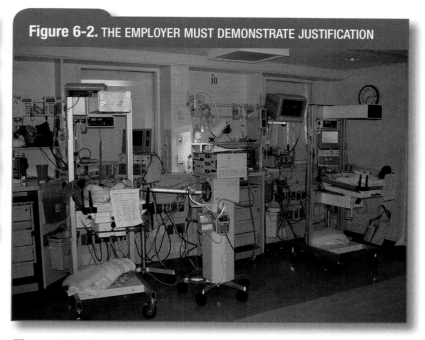

Figure 6-2. THE EMPLOYER MUST DEMONSTRATE JUSTIFICATION

Figure 6-2. The employer must demonstrate that deenergizing introduces additional hazards or increased risk or that the task to be performed is infeasible in a deenergized state due to equipment design or operational limitations before energized work is permitted. Courtesy of Michael J. Johnson-NECA.

NFPA 70E documentation requirements include those in 130.5(A) to document the arc flash risk assessment results, those in 130.5(D) as part of equipment labeling requirements, and those in 130.2(B)(2)(4) and (5) for results of the shock risk assessment and arc flash risk assessment, for example.

The sixth element required to be among the minimum requirements in an energized electrical work permit requires that the means employed to restrict access of unqualified persons be documented. Also review 130.3, 130.4(C)(1),(2),(3), and 130.7(E).

Six of the eight elements required by 130.2(B)(2) are items that are required elsewhere in *NFPA 70E*. The seventh element requires "evidence of completion of a job briefing . . ." Recall that the job briefing rule in 110.1(H) requires that the information on any energized electrical work permit be covered by the employee in charge before any job is started. These two rules work hand in hand. **See Figure 6-3.**

The eighth item is the requirement for written approval by signature authorizing energized work. It is recognition that energized work is justified with all required information in items (1) through (7) known, accounted for and documented.

The energized electrical work permit and the equipment labeling requirement in 130.5(D) are therefore among the primary provisions requiring documentation of shock and arc flash risk assessment. Thoroughly review all of these requirements for a full and complete understanding of what is required. Also review the four exemptions to the work permit in 130.2(B)(3) and the example

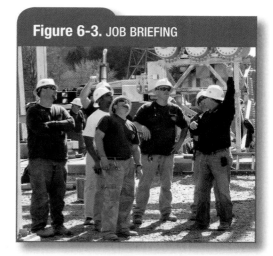

Figure 6-3. JOB BRIEFING

Figure 6-3. The information on the energized electrical work permit is required to be covered during a job briefing. Courtesy of Service Electric Company.

Figure 6-4. DETERMINING SAFETY-RELATED WORK PRACTICES

Figure 6-4. Appropriate safety-related work practices must be determined using a shock risk assessment and an arc flash risk assessment.

Figure 6-5. DETERMINING PROPER PPE FOR SHOCK PROTECTION

Figure 6-5. Determination of the personal protective equipment necessary to minimize the possibility of electric shock is one reason why a shock hazard analysis is required. *Courtesy of Salisbury by Honeywell.*

Figure 6-6. APPROACH BOUNDARIES FOR SHOCK PROTECTION FOR AC SYSTEMS

Table 130.4(D)(a) Approach Boundaries to Energized Electrical Conductors or Circuit Parts for Shock Protection for Alternating-Current Systems (All dimensions are distance from energized electrical conductor or circuit part to employee.)

(1)	(2)	(3)	(4)
	Limited Approach Boundary[b]		Restricted Approach Boundary[b]; Includes Inadvertent Movement Adder
Nominal System Voltage Range, Phase to Phase[a]	Exposed Movable Conductor[c]	Exposed Fixed Circuit Part	
<50 V	Not specified	Not specified	Not specified
50 V–150 V[d]	3.0 m (10 ft 0 in.)	1.0 m (3 ft 6 in.)	Avoid contact
151 V–750 V	3.0 m (10 ft 0 in.)	1.0 m (3 ft 6 in.)	0.3 m (1 ft 0 in.)
751 V–15 kV	3.0 m (10 ft 0 in.)	1.5 m (5 ft 0 in.)	0.7 m (2 ft 2 in.)
500 kV–550 kV	5.8 m (19 ft 0 in.)	5.8 m (19 ft 0 in.)	3.6 m (11 ft 10 in.)
765 kV–800 kV	7.2 m (23 ft 9 in.)	7.2 m (23 ft 9 in.)	4.9 m (15 ft 11 in.)

Note (1): For arc flash boundary, see 130.5(A).
Note (2): All dimensions are distance from exposed energized electrical conductors or circuit part to employee.
[a] For single-phase systems above 250V, select the range that is equal to the system's maximum phase-to-ground voltage multiplied by 1.732.
[b] See definition in Article 100 and text in 130.4(D)(2) and Informative Annex C for elaboration.
[c] *Exposed movable conductors* describes a condition in which the distance between the conductor and a person is not under the control of the person. The term is normally applied to overhead line conductors supported by poles.
[d] This includes circuits where the exposure does not exceed 120V.

Figure 6-6. The shock protection boundaries for AC systems are determined from Table 130.4(D)(a). That table is shown, in part. *Courtesy of NECA.*

energized work permit that is Figure J.1 in Informative Annex J.

WORKING WHILE EXPOSED TO ELECTRICAL HAZARDS

The title to Section 130.3 is "working while exposed to electrical hazards." The requirements here detail why safety-related work practices must be used, what they must be consistent with, and that they must be determined before exposure to electrical hazards using both shock and arc flash risk assessments. **See Figure 6-4.** These requirements also stipulate that only qualified persons can work where an electrically safe work condition has not been established. Review 130.3 in its entirety as well as 110.4(A)(1), the definition of qualified person in Article 100, and the training requirements for qualified persons in 110.2(D)(1).

APPROACH BOUNDARIES FOR SHOCK PROTECTION

The title to Section 130.4 is "approach boundaries to energized electrical conductors or circuit parts for shock protection." These requirements are divided into four categories:

- Shock risk assessment
- Shock protection boundaries

- Limited approach boundary
- Restricted approach boundary

Shock Risk Assessment

The shock risk assessment provisions in 130.4(A) state what the shock risk assessment required by 130.3 must determine. **See Figure 6-5.** Review these requirements in their entirety. In part, the shock risk assessment must determine:

- The exposure voltage
- The shock boundary requirements
- PPE for shock

The results of this shock risk assessment is among the information required to be included on an energized electrical work permit, on an equipment label, and covered during a job briefing. See 130.2(B) (2)(4) a, b, c, d, 130.5(D), and 110.1(H).

Also see 130.7 for requirements for personal and other protective equipment. Examine how the shock risk assessment provisions of 130.4(A) interface with personal protective equipment requirements of 130.7(C) and the other protective equipment requirements in 130.7(D). This includes, but is not limited to, 130.7(C)(1), 130.7(C)(7)(a),(c), 130.7(C)(10)(d)(2), 130.7(C)(14), Table 130.7(C)(14), 130.7(D)(1), 130.7(D)(1) (f), (g) ,(i), 130.7(F), and Table 130.7(F), for example.

Shock Protection Boundaries

The title to 130.4(B) is "shock protection boundaries." These rules identify the two shock protection boundaries, where they are applicable, that Table 130.4(D)(a) establishes distances for various ac system voltages, and that Table 130.4(D)(b) is to be used for the distances associated with various dc system voltages. **See Figures 6-6 and 6-7.** Review the requirements of 130.4(B), Table 130.4(D)(a), and that Table 130.4(D)(b) in their entirety.

Review the definitions of limited approach boundary and restricted approach boundary in Article 100. An informational note points out that, in certain instances, the arc flash boundary may be a greater distance than the limited approach boundary and that these boundaries are independent of each other. The distance

Figure 6-7. APPROACH BOUNDARIES FOR SHOCK PROTECTION FOR DC SYSTEMS

Table 130.4(D)(b) Approach Boundaries to Energized Electrical Conductors or Circuit Parts for Shock Protection, Direct-Current Voltage Systems

(1)	(2)	(3)	(4)
	Limited Approach Boundary		Restricted Approach Boundary; Includes Inadvertent Movement Adder
Nominal Potential Difference	Exposed Movable Conductor*	Exposed Fixed Circuit Part	
<100 V	Not specified	Not specified	Not specified
100 V–300 V	3.0 m (10 ft 0 in.)	1.0 m (3 ft 6 in.)	Avoid contact
301 V–1 kV	3.0 m (10 ft 0 in.)	1.0 m (3 ft 6 in.)	0.3 m (1 ft 0 in.)
1.1 kV–5 kV	3.0 m (10 ft 0 in.)	1.5 m (5 ft 0 in.)	0.5 m (1 ft 5 in.)
5 kV–15 kV	3.0 m (10 ft 0 in.)	1.5 m (5 ft 0 in.)	0.7 m (2 ft 2 in.)
250.1 kV–500 kV	6.0 m (20 ft 0 in.)	6.0 m (20 ft 0 in.)	3.5 m (11 ft 6 in.)
500.1 kV–800 kV	8.0 m (26 ft 0 in.)	8.0 m (26 ft 0 in.)	5.0 m (16 ft 5 in.)

Note: All dimensions are distance from exposed energized electrical conductors or circuit parts to worker.
* *Exposed movable conductor* describes a condition in which the distance between the conductor and a person is not under the control of the person. The term is normally applied to overhead line conductors supported by poles.

Figure 6-7. *The shock protection boundaries for DC systems are determined from Table 130.4(D)(b). That table is shown, in part.* Courtesy of NECA.

associated with each shock boundary is determined by voltage. Conversely, the arc flash boundary distance is not a fixed distance and is not based solely on voltage. See 130.5(B), for example.

Also recall that these boundary distances are among the information required to be included on an energized electrical work permit. See 130.2(B)(2) (4)b, c. Therefore it is also required to be covered in any job briefing in accordance with 110.1(H).

Limited Approach Boundary

The title to 130.4(C) is "limited approach boundary." These rules are broken down into three subcategories:

1. Approach by unqualified persons
2. Working at or close to the limited approach boundary
3. Entering the limited approach boundary

Review the requirements of 130.4(C) in their entirety. Among other things, these limited approach boundary rules address:

- an unqualified person's approach near the limited approach boundary,
- what the designated person in charge must advise unqualified person(s) of and where,
- what must occur if there is a need for an unqualified person to cross the limited approach boundary, and
- that an unqualified person is not permitted to cross the restricted approach boundary

Pay particular attention to the limited approach boundary and its relevance to unqualified persons. Recall that the requirements of 130.4(B) reference Table 130.4(D)(a) and Table 130.4(D)(b). Table 130.4(D)(a) is intended for use with AC systems. Table 130.4(D)(b) is intended for use with DC voltage systems. Note that both tables have four columns numbered (1) to (4) from left to right. Column (1) addresses nominal system voltage and nominal potential difference, respectively. Columns (2) through (4) address boundary distances. Also pay attention to the notes attached to the tables. For example, Note c to Table 130.4(D)(a) and the "*" to Table 130.4(D)(b) clarify the application of columns (2) and (3) and indicate which of these columns would be used to determine the limited approach boundary for a particular application.

Note the limited approach distance associated with a particular voltage after determining if it is the column (2) distance for an "exposed movable conductor" or the column (3) distance for an "exposed fixed circuit part." This limited approach boundary distance is among the information required to be included on an energized electrical work permit. See 130.2(B)(2)(4)b. It is also information required to be covered during any job briefing per 110.1(H).

Restricted Approach Boundary

The title to 130.4(D) is "restricted approach boundary." While the requirements in 130.4(C) primarily addressed the limited approach boundary and its

relationship to unqualified persons, note that the requirements in 130.3(D) primarily address the restricted approach boundary and its relationship to qualified persons. These rules are broken down into three subcategories. They are relatively complex in structure and application. Review these requirements in their entirety for a complete understanding of the three conditions whereby a qualified person could approach or take any conductive object closer to exposed energized conductors or circuit parts operating at 50 volts or more than the restricted approach boundary.

Note the restricted approach distance associated with a particular voltage from column (4) in Table 130.4(D)(a) for use with AC systems and Table 130.4(D)(b) for use with DC voltage systems. This restricted approach boundary distance is among the information required to be included on an energized electrical work permit. See 130.2(B)(2)(4)c.

ARC FLASH RISK ASSESSMENT

The title to Section 130.5 is "arc flash risk assessment." These rules, as well as those in 130.3, require that an arc flash risk assessment be performed. The assessment provisions of 130.5 require, in part, that:

(1) Determination be made whether and arc flash hazard exists,
(2) The assessment be updated as necessary with periodic review not to exceed 5 years, and
(3) The design of the overcurrent protective device, its opening time, and condition of maintenance be considered.

Review the requirements of 130.5 in their entirety. Also review the Article 100 definition of "arc flash hazard" and its two informational notes in their entirety. Closely review the informational note guidance related to the possibility of an arc flash hazard existing when parts are exposed or when they are within equipment in a guarded or enclosed condition.

Note that if the arc flash risk assessment indicates that an arc flash hazard exists, then 130.5(1) requires determination of appropriate safety-related work practices, the arc flash boundary, and the PPE to be used within the arc flash boundary.

The requirements of 130.5 are further divided into four headings:

- Documentation
- Arc flash boundary
- Arc flash PPE
- Equipment labeling

Documentation

The heading of 130.5(A) is "documentation." Here it is required that the results of the arc flash risk assessment be documented. Although not specifically mentioned where documentation is to be made, recall that this is among the information required to be included on an energized electrical work permit and therefore covered in any job briefing. See 130.2(B)(2)(3),(5),(7) and 110.1(H), for example. Also see 130.5(D), equipment labeling, for other examples of arc flash risk assessment documentation requirements. In addition to what is required to be included on the label, note that "the method of calculating and the data to support the information for the label shall be documented."

Arc Flash Boundary

The heading of 130.5(B) is "arc flash boundary." Review the definition of "arc flash boundary" and its informational note in Article 100. Also review the Article 100 definition of "incident energy" noting that incident energy "is typically expressed in calories per square centimeter (cal/cm^2).

Per 130.5(B)(1) and (2), the arc flash boundary can be determined in one of two ways.

1. The distance at which the incident energy equals 1.2 cal/cm^2.
2. By Table 130.7(C)(15)(A)(b) or Table 130.7(C)(15)(B), when the requirements of these tables apply

The requirements of 130.5(B)(1)(1) offer one of two options for determining the arc flash boundary. In accordance with this requirement, the arc flash boundary is the distance where the incident energy equals 1.2 cal/cm^2. An informational note points to Informative An-

nex D for non-mandatory information on estimating the arc flash boundary. See Table D.1, limitation of calculation methods, D.2.1, and D.4.5, for example.

The requirements of 130.5(B)(2)(2) offer the other of the two options for determining the arc flash boundary. In accordance with this requirement, the arc flash boundary shall be permitted to be determined by Table 130.7(C)(15)(A)(b) or Table 130.7(C)(15)(B), when the requirements of these tables apply.

Note that an arc flash boundary distance is listed in the "Arc-Flash Boundary" column in each of these tables. However, per 130.5(B)(2)(2), this distance is only permitted to be used "when the requirements of these tables apply." See 130.7(C)(15)(A), 130.7(C)(15)(B), Table 130.7(C)(15)(A)(a), Table 130.7(C)(15)(A)(b), and Table 130.7(C)(15)(B) for examples of the requirements that must be met to use the arc flash boundary listed in those tables.

Regardless of which of these two methods is used to determine the arc flash boundary, the arc flash boundary must be documented per 130.5(A). It must be on the energized electrical work permit per 130.2(B)(2)(5)c. It must be on a label on the equipment per 130.5(D)(2). The method of calculating and the data to support the information for the label must be documented per 130.5(D). See D.2.1 and D.4.5 for examples of "the method of calculating and the data to support the information for the label" that must be documented.

Arc Flash PPE

The heading of 130.5(C) is "Arc Flash PPE." This heading addresses the third of four components required as part of an

OSHA Tip

1910.335(a)(1)(i)
Employees working in areas where there are potential electrical hazards shall be provided with, and shall use, electrical protective equipment that is appropriate for the specific parts of the body to be protected and for the work to be performed.

arc flash risk assessment. Note that 130.5(C) identifies two methods for the selection of arc flash PPE:

1. Incident energy analysis method
2. Arc flash PPE categories method

Even though either method is permitted to be used for the selection of arc flash PPE, 130.5(C) prohibits both methods from being used on the same piece of equipment. Review 130.5(C) in its entirety, including the text that prohibits using the results of an incident energy analysis to specify an arc flash PPE category.

Incident Energy Analysis Method. One of the two methods to be used for the selection of arc flash PPE is covered in 130.5(C)(1). It is the "incident energy analysis method." Review the Article 100 definition of incident energy analysis. Thoroughly review all of the requirements of 130.5(C)(1) for specific details on:

- What the incident energy exposure level must be based on,
- What arc-rated clothing and other PPE used must be based on, and
- How to address protecting parts of the body that are closer than the distance at which incident energy was determined.

Also see the informational note to 130.5(C)(1). Review the referenced information in Informative Annex D for estimating incident energy and Table H.3(b) in Informative Annex H for information on the selection of PPE. Also see Table H.3(a) and Table H.3(b). These references may be useful in meeting the provisions of 130.7(C)(1) through (14) and 130.5(C)(1) where it is required that ". . . PPE shall be used by the employee based on the incident energy exposure . . ."

Use the results of this analysis to select arc-rated clothing and other PPE as required by 130.5(C)(1). Document this information on any energized electrical work permit per 130.2(B)(2)(5) and cover this information during any job briefing as required in 110.1(H). Also include this information on an equipment label and document the method of calculating and data to support the information on the label in accordance with

the provisions of 130.5(D). Recall the provisions in 130.1 indicating that all of the requirements of Article 130 apply, including when this method is used.

Arc Flash PPE Categories Method. The second of the two methods permitted to be used for the selection of arc flash PPE is covered in 130.5(C)(2). It is the "arc flash PPE categories method." It states that that the requirements of 130.7(C)(15) and 130.7(C)(16) must be met to select arc flash PPE using this method. Review these two sections thoroughly to understand the numerous conditions that must be met to use this method. These include, but are not limited to, the task, available short-circuit current, fault clearing times, and working distance.

Apply the requirements of 130.7(C)(15) and 130.7(C)(16) for the selection of arc flash PPE as provided for in 130.5(C)(2). Document the information derived from the use of this method on any energized electrical work permit per 130.2(B)(2)(5) and cover this information during any job briefing as required in 110.1(H). Also include this information on an equipment label and document the method of calculating and data to support the information on the label in accordance with the provisions of 130.5(D). As is the case with the application of any requirement, be sure to account for all applicable *NFPA 70E* requirements. Consider the provisions in 130.1 stating that all of the requirements of Article 130 apply, including when this method is used. Recall the provisions of 130.5 requiring, in part, that the arc flash risk assessment shall "take into consideration the design of the overcurrent protective device and its opening time, including its condition of maintenance."

Equipment Labeling

The heading of 130.5(D) is "Equipment Labeling." This heading addresses the fourth of four components required as part of an arc flash risk assessment. **See Figure 6-8.** Note that 130.5(D) does not require a field-marked label on all electrical equipment under all conditions of installation and use. The rule provides examples of, and the conditions under

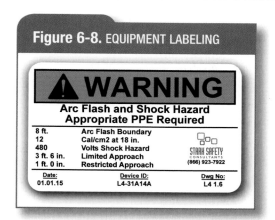

Figure 6-8. EQUIPMENT LABELING

Figure 6-8. Equipment labeling is a component of an arc flash risk assessment in NFPA 70E. Many of the labels installed on equipment provide information beyond the minimum required by 130.5(D) such as the one shown here. *Courtesy of Stark Safety Consultants.*

which, electrical equipment must be field-marked with a label. Review these requirements in their entirety. Pay particular to attention to the "likely to require" provisions and the examples of types of electrical equipment requiring a field-installed label by the use of "such as" in conjunction with this list of equipment.

The requirements of 130.5(D)(1),(2) and (3) specify what information must field-marked on any required equipment label. This includes, in part, (1) the nominal system voltage, (2) the arc flash boundary, and (3) at least one of the following:

- The available incident energy and the corresponding working distance, or the arc flash PPE category for the equipment, but not both
- The minimum arc rating of clothing
- Site-specific level of PPE

Review these requirements regarding what must be on the label in their entirety. Review the Article 100 definitions of incident energy and arc rating and consider their application in this equipment labeling requirement. "Site-specific level of PPE" was included as an additional option to (D)(1), incident energy or arc flash PPE category, and (D)(2), arc rating, for those employers or workplaces that specify their level of PPE in ways other than those identified in 130.5(D)(1) and (D)(2).

Review the exception to the equipment labeling requirement, including the date noted and what information the label must contain for the exception to apply.

Also carefully review the last two paragraphs in 130.5 in their entirety. Required here, in part, is:

- Documentation of the method of calculating and the data to support the information for the label
- Updating of the label where a review of the arc flash risk assessment identifies a change that makes the information on the label inaccurate
- Recognition that the owner of the electrical equipment is responsible for the field-marked label documentation, installation, and maintenance.

OTHER PRECAUTIONS FOR PERSONNEL ACTIVITIES

The next category of requirements related to work involving electrical hazards addresses "other precautions for personnel activities." These provisions are broken down into fourteen categories of requirements:

- Alertness
- Blind Reaching
- Illumination
- Conductive Articles Being Worn
- Conductive Materials, Tools, and Equipment Being Handled
- Confined or Enclosed Work Spaces
- Doors and Hinged Panels
- Clear Spaces
- Housekeeping Duties
- Occasional Use of Flammable Materials
- Anticipating Failure
- Routine Opening and Closing of Circuits
- Reclosing Circuits After Protective Device Operation
- Safety Interlocks

A number of these topics are similar to what OSHA requires in Subpart S of 29 CFR Part 1910. Each of these topics is an important consideration that must be carefully reviewed in their entirety, with most—if not all—of these issues likely to be included in an electrical safety program or procedures that are based on *NFPA 70E.*

An abbreviated look at a number of these 130.6 provisions follows. Again, these, like all *NFPA 70E* requirements, must be reviewed in their entirety for a full and complete understanding:

- Instruction must generally be provided to be alert at all times for work inside the limited approach boundary.
- Working inside the limited approach boundary is generally prohibited while alertness is recognizably impaired.
- Illumination must be provided that enables work to be performed safely prior to employee entry into spaces containing electrical hazards.
- Conductive articles cannot be worn within the restricted approach boundary or where they present an electrical contact hazard.
- Conductive materials, tools, and equipment must be handled in a manner that prevents accidental contact with energized electrical conductors or circuit parts.
- Conductive materials must not approach exposed energized electrical conductors or circuit parts closer than that permitted by 130.2.

- Protective shields, protective barriers, or insulating materials must generally be provided and used to avoid the effects of electrical hazards and inadvertent contact with exposed energized parts before entry into a confined or enclosed space.
- Required working space must not be used for storage.

PERSONAL AND OTHER PROTECTIVE EQUIPMENT

The requirements related to personal and other protective equipment are broken down into two broad categories: personal protective equipment and other protective equipment. PPE, such as arc flash protective equipment, shock protection, and head, face, eye, and hearing protection, is covered in 130.7(C). **See Figure 6-9.** Other protective equipment, such as insulated tools, protective shields, and rubber insulating equipment used for protection from accidental contact, is covered in 130.7(D). **See Figure 6-10.**

Keep the rules of 130.1 in mind as personal and other protective equipment requirements are considered. All requirements in Article 130 are applicable whether an incident energy analysis is completed or if the "arc flash PPE categories method" is used.

The requirements of 130.7 begin with two topic areas that apply to both the personal protective and other protective equipment requirements. The two topic areas include general requirements and provisions addressing care of equipment. The general requirement in 130.7(A) requiring, in part, that PPE be provided and used is similar to what OSHA requires in 1910.335(a)(1)(i). There are three informational notes associated with the general requirements, and one such note associated with care of equipment. Informational Note No. 1 to 130.7(A) advises, in part, that even when protection is selected in accordance with the PPE requirements of 130.7, some situations could result in burns to the skin, although burn injury should be reduced and survivable.

Figure 6-9. ARC FLASH PROTECTIVE EQUIPMENT

Figure 6-9. Personal protective equipment includes arc flash protective equipment. Courtesy of Salisbury by Honeywell.

Section 130.7(B) addresses the need for protective equipment to be:
- Maintained in a safe, reliable condition.
- Visually inspected before each use.
- Stored in a manner to prevent damage.

Personal Protective Equipment (PPE)

The Article 130 requirements for PPE are primarily covered in 130.7(C). These PPE requirements must be reviewed and applied in their entirety. They are broken down into the following 16 categories:
- General
- Movement and Visibility
- Head, Face, Neck, and Chin (Head Area) Protection
- Eye Protection
- Hearing Protection
- Body Protection
- Hand and Arm Protection
- Foot Protection
- Factors in Selection of Protective Clothing
- Arc Flash Protective Equipment
- Clothing Material Characteristics
- Clothing and Other Apparel Not Permitted
- Care and Maintenance of Arc-Rated Clothing and Arc-Rated Arc Flash Suits
- Standards for Personal Protective Equipment (PPE)
- Selection of Personal Protective Equipment (PPE) When Required for Various Tasks
- Protective Clothing and Personal Protective Equipment (PPE)

In addition to the general requirements of 130.7(A) that cover both personal and other protective equipment, certain general requirements apply specifically to personal protective equipment. The requirements in 130.7(C)(1) address both shock and arc flash protective equipment. **See Figure 6-11.**

When an employee is working within the restricted approach boundary, he or she must wear PPE in accordance with 130.4. When an employee is working

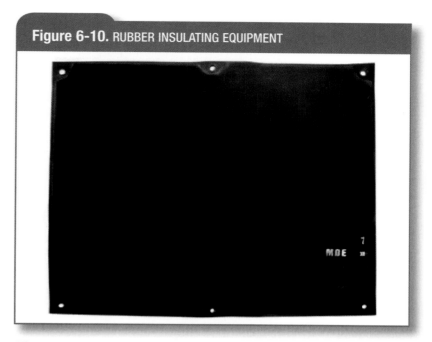

Figure 6-10. RUBBER INSULATING EQUIPMENT

Figure 6-10. Other protective equipment includes rubber insulating equipment. Courtesy of Salisbury by Honeywell.

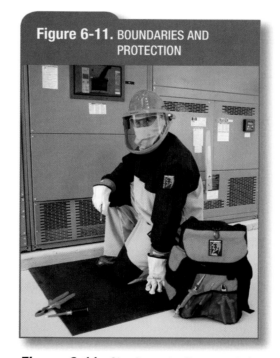

Figure 6-11. BOUNDARIES AND PROTECTION

Figure 6-11. Shock protection must be worn within the restricted approach boundary, and arc flash protection must be worn within the arc flash boundary. Courtesy of Salisbury by Honeywell.

within the arc flash boundary, he or she is required to wear protective clothing and other personal protective equipment in accordance with 130.5. All parts of the body inside the arc flash boundary must be protected. Movement and visibility is covered in 130.7(C)(2). When arc-rated clothing is worn as protection, it is required to cover all ignitable clothing. The arc-rated clothing worn must also allow for movement and visibility. Also review 130.7(C)(2) and 130.7(C)(9)(d),(e),(f).

Protection. The PPE requirements of 130.7(C) provide a number of protection requirements for both general and specific parts of the body. Keep in mind the provisions of 130.1 as the following abbreviated protection requirements are explored. These and other requirements may be applicable even where the "arc flash PPE categories method" is used rather than the "incident energy analysis method." Review 130.1, 130.5, and 130.5(C).

Note that the arc flash protective equipment requirements of 130.7(C) (10) may modify or supplement one or more of the requirements of 130.7(C) as indicated by the informational note to the head, face, neck, chin, and eye protection requirements in 130.7(C)(3), for example. **See Figure 6-12.**

Section 130.7(C) is divided into 16 subdivisions: 130.7(C)(1) through (14) are used to select PPE to fulfill shock and arc flash PPE needs. See also Table H.3(a) and Table H.3(b) in Informative Annex H. 130.7(C)(15) addresses identifying when arc flash PPE is required and the selection of PPE when required for various tasks in lieu of the incident energy analysis method of 130.5(C)(1). 130.7(C) (16) addresses protective clothing and PPE once the arc flash PPE category has been identified from Table 130.7(C)(15)(A)(b) and Table 130.7(C)(15)(B). Recall that, per 130.1, 130.7(C)(1) through (14) are also applicable to the provisions of 130.7(C)(15), Table 130.7(C)(15)(A)(a), Table 130.7(C)(15)(A)(b), Table 130.7(C)(15)(B), 130.7(C)(16), and Table 130.7(C)(16), for example.

Figure 6-12. MODIFIED OR SUPPLEMENTED REQUIREMENTS

Figure 6-12. *Specific arc flash protective equipment requirements may modify or supplement other PPE requirements for specific parts of the body.* Courtesy of Salisbury by Honeywell.

An overview of these PPE requirements follows. Again, these, like all requirements in *NFPA 70E*, need to be reviewed and applied in their entirety. Note that Informative Annex H provides additional nonmandatory information that may help in the understanding and application of these and other requirements.

These PPE requirements begin with those related to head, face, neck, and chin (head area) protection in 130.7(C) (3). In addition to the need for nonconductive head protection and protective equipment for the face, neck, and chin, as necessary, hairnets and beard nets, where used, must be arc rated when working within the arc flash boundary. Eye protection is addressed in 130.7(C) (4). It is generally required in addition to face protection and must be worn whenever there is danger of injury from electric arcs, flashes, or from flying objects resulting from electrical explosion.

Hearing protection is required in each of the four arc flash PPE categories in Table 130.7(C)(16). Hearing protection is also addressed in 130.7(C)(5) and is required when working inside the arc flash boundary.

The "body protection" provisions of 130.7(C) (6) require arc-rated clothing wherever there is possible exposure to an electric arc flash above 1.2 cal/cm². Body protection includes all parts of the body as clarified in the committee statement on Proposal 70E-271 for the 2009 edition of *NFPA 70E*. That proposal sought to replace the word "body" with the word "torso." The *NFPA 70E* Technical Committee rejected that recommendation with a unanimous 24-0 vote with the following response:

Committee Statement: "Removing the word 'body' and replacing it with the word 'torso' could mislead the user of the standard by implying that other parts of the body need not be protected. The committee concludes that this section is intended to address all parts of the body."

Hand and arm protection requirements are set forth in 130.7(C)(7). **See Figure 6-13.** These hand and arm protection provisions are divided into the following categories:

- Shock protection
- Arc flash protection
- Maintenance and use

The rules in 130.7(C)(7)(a) for shock protection require that rubber insulating gloves be rated for voltage exposure. Requirements are also provided on when rubber insulating gloves with leather protectors must be worn as well as when they must be worn in conjunction with rubber insulating sleeves. In addition to a complete examination of these rules, review the exception addressing when it is necessary to wear rubber insulating gloves without leather protectors. Also review the standards on protective equipment in Table 130.7(C)(14). Standards for rubber insulating gloves, insulating sleeves, and leather protector gloves are among the standards listed here. Per

Figure 6-13. RUBBER INSULATING SLEEVES

Figure 6-13. Rubber insulating sleeves must be worn in addition to insulating gloves with leather protectors where there is a danger of electric shock due to contact with energized electrical conductors or circuit parts.
Courtesy of Salisbury by Honeywell.

130.7(C)(14), PPE must meet the standards in this table.

The rules in 130.7(C)(7)(b) for arc flash protection state that hand and arm protection must be worn where there is possible exposure to arc flash burn. Review this requirement in its entirety and note the additional references to 130.7(C) (10)(d) and 130.7(C)(6) for compliance with hand and arm arc flash protection provisions.

The rules in 130.7(C)(7)(c) address maintenance and use of hand and arm protection. How electrical protective equipment must be maintained, when and how insulating equipment must be tested and inspected for damage, and the test voltages and the maximum intervals between tests are among the requirements set forth here. Note what is covered in Table 130.7(C)(7)(c). In addition to examining the requirements of 130.7(C)(7)(c) in their entirety, also review and explore the informational note

OSHA Tip

The regulations for electrical protective equipment in 1926.97 are nearly identical to those for electrical protective equipment in 1910.137. Part 1926 provides Safety and Health Regulations for Construction. Those in part 1910 are Occupational Safety and Health Standards.

Figure 6-14. ELECTRICAL PROTECTIVE DEVICES

Part Number	1910
Part Title	Occupational Safety and Health Standards
Subpart	I
Subpart Title	Personal Protective Equipment
Standard Number	1910.137
Title	Electrical protective devices.

Figure 6-14. OSHA's requirements in 1910.137 address electrical protective devices. These provisions are located within Subpart I, Personal Protective Equipment.

references to OSHA 1910.137 and ASTM F496. **See Figure 6-14.**

The title to 130.7(C)(8) is "foot protection." Note that, while "foot protection" is covered here in 130.7(C)(8), these requirements address protection against step and touch potential, the need for dielectric footwear, and limitations of insulated soles. An informational note provides insight into a standard addressing electrical hazard footwear. Review the rules in 130.7(C)(8) in their entirety and note that this requirement does not address foot protection as it relates to protection against arc flash. Also review 130.7(C)(10)(e), 130.7(A), 130.7(C)(1), and Table 130.7(C)(16) as foot protection requirements are considered.

Factors in Selection of Protective Clothing. The 130.7(C)(9) requirements address factors in the selection of protective clothing. The rules that apply generally require several things. First, similar to what is required in 130.7(C)(2), flammable clothing, as well as all associated parts of the body, must be covered when arc-rated clothing is required while allowing for movement and visibility. Second, similar to what is required per 130.7(C)(1), clothing and equipment that protect against shock and arc flash

hazards must be used. Additional requirements in 130.7(C)(9) clarify whether garments without an arc rating can increase an arc rating and if clothing and equipment can be worn alone or be used with flammable, nonmelting apparel. An informational note discusses examples of protective clothing, including arc-rated rainwear.

Review the general requirements in 130.7(C)(9) in their entirety. Note that the "factors in selection of protective clothing" provisions are further addressed in six additional categories:
* Layering
* Outer layers
* Underlayers
* Coverage
* Fit
* Interference

Requirements in the first three of these six subcategories address "layers and layering." These are covered in 130.7(C)(9) (a),(b), and (c) and detail the use of nonmelting flammable fiber garments, the need for arc-rated garments worn as outer layers, and the prohibition of meltable fibers used as underlayers next to the skin. **See Figure 6-15.** A number of these considerations include:
* If nonmelting, flammable fiber garments are used as underlayers, the

Figure 6-15. OUTER LAYERS

Figure 6-15. Garments worn as outer layers over arc-rated clothing must be made from arc-rated material. Courtesy of Salisbury by Honeywell.

system arc rating must be sufficient to prevent breakopen of the inner-most arc-rated layer at the expected arc exposure incident energy level to prevent ignition of flammable underlayers.

- Garments that are not arc rated cannot increase the arc rating of a garment or of a clothing system.
- Garments worn as outer layers over arc-rated clothing, such as jackets or rainwear, must also be made from arc-rated material.
- Meltable fibers are generally not permitted in fabric underlayers (underwear) next to the skin.
- An incidental amount of elastic used on nonmelting fabric under-wear or socks shall be permitted.

Finally, coverage, fit, and interference are additional topics that are included in the requirements to be considered as factors in the selection of protective clothing. **See Figure 6-16.** A number of these considerations in 130.7(C)(9)(d), (e), and (f) include:

- Tight-fitting clothing must be avoided.
- Arc-rated apparel must fit properly so as not to interfere with the task.
- Clothing must cover potentially exposed areas as completely as possible.
- Shirt and coverall sleeves must be fastened at the wrists, shirts must be tucked into pants, and shirts, coveralls, and jackets must be closed at the neck.
- The garment selected shall result in the least interference with the task but still provide the necessary protection.

Review all of the requirements, exceptions, and informational notes related to these six subcategories in their entirety as well. Additional requirements that apply to "factors in selection of protective clothing" include those set forth in 130.7(C)(11), (12), and (14) and will therefore be covered next.

Clothing Material Characteristics. *NFPA 70E* Article 130 also provides requirements addressing the clothing material characteristics; these are primarily located

Figure 6-16. SHIRT AND COVERALL SLEEVES

Figure 6-16. *Shirt and coverall sleeves must be fastened at the wrist.*

in 130.7(C)(11). If the protective clothing is arc rated, it must meet the requirements of 130.7(C)(12) and 130.7(C)(14). Table 130.7(C)(14) identifies the standards that PPE, such as arc-rated apparel, rubber insulating gloves, leather protectors, and arc-rated face protective products, must conform to. Three informational notes provide information related to the application of this provision.

130.7(C)(11) further requires that clothing incorporating fabrics, zipper tapes, and findings made from flammable synthetic materials that melt at temperatures below 315°C (600°F), such as acetate, acrylic, nylon, polyester, polyethylene, polypropylene, and spandex, either alone or in blends, must not be used. An informational note to this requirement provides insight into this provision.

Review all of these requirements, exceptions, and informational notes in their entirety, including all cross references and tables for standards for PPE. Also consider the provisions of 130.1 whereby all of the requirements of Article 130 apply whether an incident analysis method or the arc flash PPE category method is used to select arc flash PPE.

Types of Clothing and Other Apparel Not Permitted. The PPE requirements given in 130.7(C)(12) address the types of clothing and other apparel that are not permitted. Note that these requirements reference 130.7(C)(11), which in turn refers back to 130.7(C) (12) as well as to Table 130.7(C)(14). Table 130.7(C) (14) identifies the standards that PPE must conform to.

Clothing and other apparel (such as hard hat liners and hair nets) made from materials that do not meet the requirements of 130.7(C)(11) regarding melting, or made from materials that do not meet the flammability requirements, are generally not be permitted to be worn.

An informational note and two exceptions to the requirements of 130.7(C) (12) are provided. Exception No. 1 to 130.7(C)(12) conditionally allows non-melting, flammable clothing to be used as underlayers when it is worn under arc-rated clothing.

As always, review all of these requirements, exceptions, and informational notes in their entirety and consider the application of the requirements of 130.1 regardless of the method is used to select arc flash PPE.

Arc Flash Protective Equipment. The arc flash protective equipment requirements are addressed in 130.7(C)(10) and are divided into five categories. In addition to detailing unique and specific arc flash protective equipment requirements, these rules supplement or modify other requirements in 130.7(C), such as indicated in the informational note to 130.7(C)(3). The five major topics addressed by 130.7(C)(10) are as follows:

- Arc flash suits
- Head protection
- Face protection
- Hand protection
- Foot protection

An abbreviated look at these requirements follows. As these requirements are reviewed in full, consider their application throughout Article 130, such as to Table 130.7(C)(16), as indicated in 130.1. Recall too how these "arc flash protective equipment" requirements modify or supplement other rules in Article 130. See 130.7(C)(3), (4), (6), (7), and (8), for example.

ARC FLASH SUITS. The first of the five major topics is addressed in 130.7(C) (10)(a), arc flash suits. **See Figure 6-17.** By definition, an arc flash suit must cover the entire body except for the hands and feet. See article for the defined term "arc flash suit." A number of arc flash suit considerations, in part, include:

- Design must permit easy and rapid removal.
- The entire arc flash suit must have an arc rating suitable for the arc flash exposure.
- Covering or construction of air hoses and pump housing when exterior air is supplied into the hood.

HEAD PROTECTION. The second of the five major topics addressed in 130.7(C) (10) is head protection. **See Figure 6-18.**

In addition to meeting other head protection requirements, such as those in 130.7(C)(3), 130.7(C) (14), and Table 130.7(C)(14), head protection must meet head protection requirements for arc flash protective equipment in 130.7(C)(10)(b). In part, these considerations include:

- When an an arc-rated balaclava is required to be used with an arc-rated faceshield.
- When an an arc-rated hood must be used instead of an arc-rated faceshield and balaclava combination.

Figure 6-17. ARC FLASH SUIT

Figure 6-17. The entire arc flash suit, including the hood's face shield, must have an arc rating that is suitable for the arc flash exposure. Courtesy of Salisbury by Honeywell.

Figure 6-18. HEAD PROTECTION

Figure 6-18. An arc-rated hood or an arc-rated balaclava in conjunction with an arc-rated face shield is required when the back of the head is within the arc flash boundary. Courtesy of Salisbury by Honeywell.

Again, recall that 130.1 states that all requirements of Article 130 apply whether an "incident energy analysis method" or the "arc flash PPE category method" is used. Apply the head protection requirements for arc flash protective equipment, as applicable, as protective clothing and equipment is determined by Table 130.7(C)(16), for example.

FACE PROTECTION. The third of the five major topics addressed in 130.7(C) (10) is face protection. In addition to meeting other face protection requirements, such as those in 130.7(C) (3), 130.7(C)(14), and Table 130.7(C)(14), face protection must meet the specific face protection requirement for arc flash protective equipment provided in 130.7(C) (10)(c). In particular, face shields with an adequate arc rating must be worn, and eye protection must be worn even when face protection is worn. **See Figure 6-19.** Face shields must also incorporate wrap-around guarding that protects the face, chin, forehead, ears, and neck area. An informational note to these requirements provides insight into this provision, including the potential need for additional illumination, as these shields are tinted and can reduce visual acuity and color perception.

Figure 6-19. EYE AND FACE PROTECTION

Figure 6-19. Eye protection is always required to be worn under face shields or hoods. Courtesy of Salisbury by Honeywell.

HAND PROTECTION. The fourth of the five major topics addressed in 130.7(C)(10) is hand protection. In addition to meeting other hand protection requirements, such as those in 130.7(C)(7), 130.7(C)(14), and Table 130.7(C) (14), hand protection must meet the requirements for arc flash protective equipment, when necessary. Heavy-duty leather gloves or arc-rated gloves must be worn when required for arc flash protection. Arc-rated gloves and heavy-duty leather gloves are required to provide arc flash protection for the hands. Where insulating rubber gloves are used for shock protection, leather protectors must be worn over the rubber gloves. Additional information is provided in the two informational notes. This includes the thickness of leather that would qualify leather gloves as heavy duty and insight into the protection provided by leather protectors worn over rubber insulating gloves for arc flash protection.

FOOT PROTECTION. The fifth of these five major "arc flash protective equipment" topics addressed in 130.7(C)(10)(e) covers foot protection. In addition to meeting other foot protection requirements, such as those in 130.7(C)(8), 130.7(C)(14), and Table 130.7(C)(14), foot protection must be in the form of heavy-duty leather footwear where arc flash exposures exceed 4 cal/cm^2.

As 130.1 is considered, compare this requirement with the foot protection required by Table 130.7(C)(16) where leather footwear is required for PPE Categories 1, 2, 3, and 4 per this table. Even though there is no method recognized in the requirements of *NFPA 70E* to equate an incident energy calculation with an arc flash PPE category, it is worth noting that Table 130.7(C)(16) requires "leather footwear" for all PPE categories (including for PPE categories 2, 3 and 4), while the provisions of 130.7(C)(10)(e) require heavy-duty leather footwear above 4 cal/cm^2.

Care and Maintenance of Arc-Rated Clothing and Arc-Rated Arc Flash Suits. Several requirements cover the care and maintenance of arc-rated clothing and arc flash suits. **See Figure 6-20.** These rules are primarily located in 130.7(C)(13) and are divided into four categories:
- Inspection
- Manufacturer's Instructions
- Storage
- Cleaning, Repairing, and Affixing Items

Compliance with ASTM F 1506 as provided for by Table 130.7(C)(14), and the following summary of provisions are among what is required by these rules:
- Apparel must be inspected before each use.
- Garments that are contaminated or damaged to the extent that their protective qualities are impaired cannot be used.

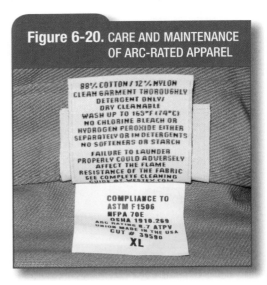

Figure 6-20. CARE AND MAINTENANCE OF ARC-RATED APPAREL

Figure 6-20. The garment manufacturer's instructions for care and maintenance of arc-rated apparel must be followed to avoid loss of protection.

- Manufacturers' instructions for care, maintenance, and cleaning must be followed.
- The same arc-rated materials used to manufacture the clothing must be used for repair.
- Guidance in ASTM F 1506 must be followed when trim, name tags, or logos are affixed.

In addition to reviewing the two informational notes following 130.7(C)(13)(d), study the requirements in the following four categories of topics that were introduced in the list above from 130.7(C)(13). These and all related requirements must be referred to in their entirety for a full and complete understanding.

Standards for Personal Protective Equipment (PPE). Not only does *NFPA 70E* require protective equipment, but it also requires that much of this protective equipment meet specific standards. Notably, two tables specify these protective equipment standards. Other protective equipment standards, such as those for arc protective blankets, insulated hand tools, ladders, line hose, safety signs and tags, and temporary protective grounds, are identified in Table 130.7(F). Personal protective equipment must conform to the standards specified in Table 130.7(C)(14).

This includes PPE such as arc-rated apparel, eye and face protection, arc-rated face protection, fall protection, leather protectors, rubber insulating gloves, hard hats, and arc-rated rainwear, for example.

Note that non–arc-rated or flammable fabrics are not covered by a standard in Table 130.7(C)(14), even though they are permitted to be used in some cases. The informational note to 130.7(C)(14) points to 130.7(C) (11) and 130.7(C) (12) for reference.

Selection of Personal Protective Equipment Using Arc Flash PPE Categories.

One of the required outcomes of an arc flash risk assessment in accordance with 130.5 is determining the personal protective equipment that people within the arc flash boundary must use. **See Figure 6-21.** There are two methods that may be used to make this determination: the incident energy analysis method per 130.5(C)(1) or use of arc flash PPE categories per 130.5(C)(2). The require-

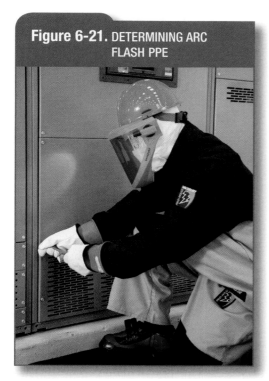

Figure 6-21. DETERMINING ARC FLASH PPE

Figure 6-21. Determination of the arc flash PPE to be used within the arc flash boundary is one reason why an arc flash risk assessment is performed. Courtesy of Salisbury by Honeywell.

ments for selecting arc flash PPE using the arc flash PPE category method will be explored here.

The requirement and informational notes in 130.7(C)(15) address protective equipment determination where arc flash PPE categories, rather than an incident energy analysis, constitute the method used to select protective equipment. The requirements for ac equipment are covered in 130.7(C)(15)(A)(b) and those for dc in 130.7(C)(15)(B).

Table 130.7(C)(15)(A)(a) is used to identify when arc flash PPE is required for both ac and dc equipment. Note that this table has three columns, a note to the table, and explanatory information related to the asterisk used in the "equipment condition" column heading. The left column of the table is where a task is listed as indicated by the heading to this column. As required in 130.7(C)(15)(A) (1) and 130.7(C)(15)(B)(1), the task to be performed must be listed in this column in order to use Table 130.7(C)(15) (A)(a). The middle column in this table is the "equipment condition*" column as indicated by its column heading. Review the explanatory information related to the "equipment condition" asterisk. This explanatory information is located at the end of table "(A)(a)" immediately after the note to the table. The right column in the table identifies whether arc flash PPE is required as indicated by the title of this table column. A "Yes" or "No" is obtained from this table based on the task listed in the left column, the equipment condition in the center column, and a yes or no - arc flash PPE required – in the right column. Be sure to review the note to this table, including the last sentence of that note. It states "where this table indicates that arc flash PPE is not required, an arc flash is not likely to occur."

When arc flash PPE is required, Table 130.7(C)(15)(A)(b) is used to determine the arc flash PPE category for ac systems and Table 130.7(C)(15)(B) is used to determine the arc flash PPE category for dc systems. Both of these tables have a number of conditions that must be met to use them to determine the appropriate protective equipment via arc flash PPE categories rather than an incident energy analysis.

Figure 6-22. ARC FLASH PPE CATEGORY FOR AC

RISK ASSESSMENT & PPE SELECTION

DETERMINING ARC FLASH HAZARD PPE CATEGORIES **ALTERNATING CURRENT**
Excerpts from Table 130.7 (C)(15)(A)(b)

Equipment	Arc Flash PPE Category	Arc-Flash Boundary	Rubber-Insualted Gloves (Minimum)
Panelboards or other equipment rated 240 V and below Parameters: Maximum of 25 kA short-circuit current available; maximum of 0.03 sec (2 cycles) fault clearing time; working distance 455 mm (18 in.)	1	485 mm (19 in.)	Class 00
Panelboards or other equipment rated >240 V and up to 600 V Parameters: Maximum of 25 kA short-circuit current available; maximum of 0.03 sec (2 cycles) fault clearing time; working distance 455 mm (18 in.)	2	900 mm (3 ft.)	Class 00 (<500V) Class 0 (>500V < 1000V)
600-V class motor control centers (MCCs) Parameters: Maximum of 65 kA short-circuit current available; maximum of 0.03 sec (2 cycles) fault clearing time; working distance 455 mm (18 in.)	2	1.5 m (5 ft)	Class 00 (<500V) Class 0 (>500V < 1000V)

Figure 6-22. *Table 130.7(C)(15)(A)(b) provides arc flash PPE categories and arc flash boundaries for AC systems when the arc flash PPE category method is used for a covered task and arc flash PPE is required. The arc flash PPE categories, arc flash boundaries, equipment and parameters shown in the excerpt from the "NECA NFPA 70E Personal Protective Equipment (PPE) Selector" are based on Table 130.7(C)(15)(A)(b).* Courtesy of NECA.

The requirements of 130.7(C)(15)(A) and 130.7(C)(15)(B) must be reviewed and applied in their entirety. If all of the conditions set in these requirements are not met then the arc flash category method cannot be used and the incident energy analysis method must be used. These requirements reference Table 130.7(C)(15)(A)(b) for ac systems and Table 130.7(C)(15)(B) for dc systems. The requirements of 130.7(C)(15)(A) and 130.7(C)(15)(B) referencing these tables stipulate that the parameters in these tables must also be met to use the arc flash PPE category method. These parameters are located under each equipment description in the "equipment" heading in these tables. The parameters for "Panelboards or other equipment rated 240 V and below" in Table 130.7(C)(15)(A)(b) are: maximum of 25 kA short-circuit available, maximum of 2 cycles fault clearing time, and a working distance of 18 inches, for example. Likewise, the parameters for work on dc systems are located under each equip-

ment description in the "equipment" heading in Table 130.7(C)(15)(B). Again, the "incident energy analysis method" must be used to select arc flash PPE if all of these parameters are not met.

In addition to thoroughly reviewing the requirements of 130.7(C)(15)(A) and (B), be sure to review the five informational notes that follow these requirements. They provide information, such as:

- How the arc flash PPE category, work tasks, and protective equipment determined by the arc flash PPE category were identified and selected.
- The conditions under which closed doors do not provide enough protection to eliminate the need for PPE.
- How "short-circuit current" as used in Table 130.7(C)(15)(B) is determined.
- What the methods for estimating the dc arc flash incident energy that was used to determine the categories for Table 130.7(C)(15)(B) are based on.

Figure 6-23. ARC FLASH PPE CATEGORY FOR DC

RISK ASSESSMENT & PPE SELECTION

DETERMINING ARC FLASH HAZARD PPE CATEGORIES **DIRECT CURRENT**
Excerpts from Table 130.7 (C)(15)(B)

Equipment	Arc Flash PPE Category	Arc-Flash Boundary	Rubber-Insualted Gloves (Minimum)
Storage batteries, dc switchboards, and other dc supply sources 100 V > Voltage < 250 V Parameters: Voltage: 250 V Maximum arc duration and working distance: 2 sec @ 455 mm (18 in.)			
Short-circuit current < 4 kA	1	900mm (3 ft.)	Class 00*
4 kA ≤ short-circuit current < 7 kA	2	1.2m (4 ft.)	Class 00*
7 kA ≤ short-circuit current < 15 kA	3	1.8m (6 ft.)	Class 00*
Storage batteries, dc switchboards, and other dc supply sources 250 V > Voltage < 600 V Parameters: Voltage: 600 V Maximum arc duration and working distance: 2 sec @ 455 mm (18 in.)			
100 V > Voltage < 250 V Parameters:	1	900mm (3 ft.)	Class 00*
Voltage: 250 V	2	1.2m (4 ft.)	Class 00*
Maximum arc duration and working distance: 2 sec @ 455 mm (18 in.)	3	1.8m (6 ft.)	Class 00*
7 kA ≤ short-circuit current < 10 kA	4	2.5m (8 ft.)	Class 00*

Figure 6-23. *Table 130.7(C)(15)(B) provides arc flash PPE categories and arc flash boundaries for DC systems when the arc flash PPE category method is used for a covered task and arc flash PPE is required. The arc flash PPE categories, arc flash boundaries, equipment and parameters shown in the excerpt from the "NECA NFPA 70E Personal Protective Equipment (PPE) Selector" are based on Table 130.7(C)(15)(B).* Courtesy of NECA.

ARC FLASH HAZARD PPE CATEGORIES FOR AC SYSTEMS. Per 130.7(C)(15)(A), when it has been determined that arc flash PPE is required for a task by Table 130.7(C)(15)(A)(a), Table 130.7(C)(15)(A)(b) is to be used to determine the arc flash PPE category for the task to be performed for ac equipment within the parameters associated with each equipment description in the "equipment" heading in the table. **See Figure 6-22.** In addition to determining the arc flash PPE category from the middle column of the table that is associated with the parameters for the equipment in the left column, the arc flash boundary is obtained from the right column of the table.

A review of this table shows that the arc flash boundary is 19 inches and the arc flash PPE category is 1 for voltage testing performed in a panelboard rated 240 V that has a maximum of 25 kA short-circuit current available, a maximum of 2 cycle clearing time, and a working distance of 18 inches, for example. This is among the information that is to be documented on an energized

electrical work permit, covered during a job briefing, and included on an equipment label. See 130.2(B)(2), 110.1(H), and 130.5(D). Recall also that 130.5(D) requires that the method of calculating and the data to support the information for the equipment label be documented. Per 130.1, also consider whether any other requirements of Article 130 apply.

In addition to reviewing this table in its entirety, review the note to Table 130.7(C)(15)(A)(b) describing the conditions where an arc flash PPE category can be reduced by one number.

ARC FLASH HAZARD PPE CATEGORIES FOR DC SYSTEMS. Per 130.7(C)(15)(B), when it has been determined that arc flash PPE is required for a task by Table 130.7(C)(15)(A)(a), Table 130.7(C)(15)(B) is to be used to determine the arc flash PPE category for the task to be performed for dc equipment within the parameters associated with each equipment description in the "equipment" heading in the table. **See Figure 6-23.** In addition to determining the arc flash PPE category from the middle column of the table that is associated with the parameters for the equipment in the left column, the arc flash boundary is obtained from the right column of the table. This is among the information that is to be documented on an energized electrical work permit, covered during a job briefing, and included on an equipment label. See 130.2(B)(2), 110.1(H), and 130.5(D). Recall also that 130.5(D) requires that the method of calculating and the data to support the information for the equipment label be documented. Per 130.1, also consider whether any other requirements of Article 130 apply.

In addition to reviewing this table in its entirety, review the note to Table 130.7(C)(15)(B) describing the additional conditions that must be met for apparel that can be expected to be exposed to electrolyte.

Protective Clothing and PPE Determination Using Arc Flash PPE Category

In accordance with 130.7(C)(16), Table 130.7(C)(16) can only be used to determine the required PPE for a task in equipment within its parameters based on an arc flash PPE category determined from Table 130.7(C)(15)(A)(b) for ac or Table 130.7(C)(15)(B) for dc. An arc flash PPE category cannot be determined through an incident energy analysis.

Once the arc flash PPE category has been identified from Table 130.7(C)(15) (A)(a) or Table 130.7(C)(15)(B), Table 130.7(C)(16) is to be used to determine the required arc flash PPE for the task. This clothing and equipment must be used when working within the arc flash boundary.

The title to Table 130.7(C)(16) is Personal Protective Equipment (PPE), the left column of this table is titled PPE Category, and the right column of this table is titled PPE. Remember that this table is only determining arc flash PPE as it is an arc flash PPE category. This protective clothing and equipment determination from Table 130.7(C)(16) is an adjunct to the shock hazard protective equipment identified in provisions such as 130.4, Table 130.4(D)(a), Table 130.4(D)(b), 130.7(C) (7), 130.7(C)(8), and 130.7(C)(10)(d)(2), for example. Per 130.1, also consider

Figure 6-24. PPE DETERMINATION VIA ARC FLASH PPE CATEGORY

RISK ASSESSMENT & PPE SELECTION

CATEGORY 2 PPE
- Arc-Rated Clothing, Minimum Arc Rating of 8cal/cm2 (see Note 1)
- Arc-rated long-sleeve shirt and pants or arc-rated coverall
- Arc-rated flash suit hood or arc-rated face shield (see Note 2) and arc-rated balaclava
- Arc-rated jacket, parka, rainwear, or hard hat liner (AN)

PROTECTIVE EQUIPMENT
- Hard hat
- Safety glasses or safety goggles (SR)
- Hearing protection (ear canal inserts)
- Heavy duty leather gloves (see Note 3)
- Leather footwear

AN: as needed (optional). AR: as required. SR: selection required.

Notes:
(1) Arc rating is defined in Article 100.
(2) Face shields are to have wrap-around guarding to protect not only the face but also the forehead, ears, and neck or, alternatively, an arc-rated arc flash suit hood is required to be worn.
(3) If rubber insulating gloves with leather protectors are used, additional leather or arc rated gloves are not required. The combination of rubber insulating gloves with leather protectors satisfies the arc flash protection requirement.

Figure 6-24. PPE is selected from Table 130.7(C)(16) based on the arc flash PPE category determined for a covered task when the arc flash PPE category method is used and arc flash PPE is required. The PPE shown in the excerpt from the "NECA NFPA 70E Personal Protective Equipment (PPE) Selector" is based on Table 130.7(C)(16). Courtesy of NECA.

other requirements in Article 130 that might apply to Table 130.7(C)(16).

The requirements and informational notes of 130.7(C)(16) detail the application of Table 130.7(C) (16). **See Figure 6-24.** Review these requirements and informational notes in their entirety, especially Informational Note No. 2. It states, in part, that, while some situations could result in burns to the skin, even with the protection described in Table 130.7(C)(16), burn injury should be reduced and survivable and that due to the explosive effect of some arc events, physical trauma injuries could occur.

Review and apply Table 130.7(C)(16) in its entirety, including associated notes. Match the PPE in the right column of the table with the PPE category in the left column of the table. Also account for any applicable requirements of Article 130 and document the arc flash PPE and arc flash PPE category on an energized electrical work permit and equipment label, as required. See 130.1, 130.2(B)(2)(5), 130.5(D), 130.7(C), and Table 130.7(C) (14), for example.

Other Protective Equipment

Personal protective equipment is covered in 130.7(C) and includes equipment such as arc flash protective equipment, shock protection, and head, face, eye, and hearing protection. Other protective equipment, such as insulated tools, protective shields, and rubber insulating equipment used for protection from accidental contact, is covered in 130.7(D). The other protective equipment required by 130.7(D) must comply with those requirements as well as the standards provided in Table 130.7(F). **See Figure 6-25.**

The requirements of 130.7(D) are broken down into nine categories of insulated tools and equipment:
- Requirements for Insulated Tools
- Fuse or Fuseholder handling
- Ropes and Handlines
- Fiberglass-Reinforced Plastic Rods
- Portable Ladders
- Protective Shields
- Rubber Insulating Equipment
- Voltage-Rated Plastic Guard Equipment
- Physical or Mechanical Barriers

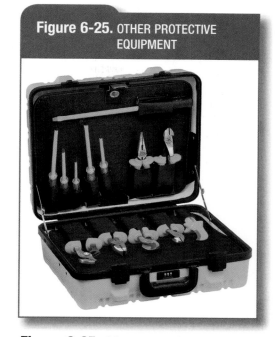

Figure 6-25. OTHER PROTECTIVE EQUIPMENT

Figure 6-25. Other protective equipment, such as insulated tools, must comply with the standards provided in Table 130.7(F). Courtesy of Klein Tools.

The requirements and informational note addressing insulated tools and equipment are located in 130.7(D)(1). The other protective equipment requirements of 130.7(D) begin with general requirements that apply to the nine categories of insulated tools and equipment. These general provisions of 130.7(D)(1), in part, state:
- Employees must use insulated tools when working inside the restricted approach boundary.
- Insulated tools must be protected from damage to the insulating material.

The specific requirements for each of the nine categories of insulated tools and equipment are contained in 130.7(D)(1) (a) through (i). Be sure to refer to *NFPA 70E* to explore the requirements in the nine categories of topics of 130.7(D)(1) that were introduced in the list above. Much of this "other protective equipment" including, but not limited to, insulated tools, portable ladders, protective shields, and rubber insulating equipment are utilized fairly often. These and all related requirements must be referred

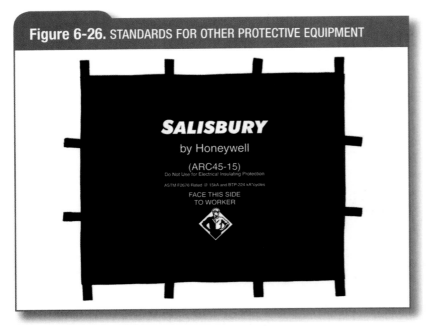

Figure 6-26. STANDARDS FOR OTHER PROTECTIVE EQUIPMENT

Figure 6-26. *Standards on Other Protective Equipment are located in Table 130.7(F).* Courtesy of Salisbury by Honeywell.

to in their entirety for a full and complete understanding.

Standards for Other Protective Equipment

NFPA 70E requires protective equipment and also mandates that much of this protective equipment meet specific standards. **See Figure 6-26.** In particular, personal protective equipment must conform to the standards specified in Table 130.7(C)(14); other protective equipment required in 130.7(D) must conform to the standards given in Table 130.7(F). Other protective equipment standards, such as those for arc protective blankets, insulated hand tools, ladders, line hose, safety signs and tags, and temporary protective grounds, are identified in Table 130.7(F).

Alerting Techniques

Alerting techniques are located in 130.7 and are therefore recognized as a form of protective equipment. These protective equipment requirements are broken down into the following categories in 130.7(E):

- Safety Signs and Tags
- Barricades
- Attendants
- Look-Alike Equipment

Safety signs and tags are covered by one of the standards in Table 130.7(F). One of the required elements of the energized work permit is "means employed to restrict access of unqualified persons from the work area." The alerting techniques identified in 130.7(E) may be among the ways to meet this requirement. Refer to *NFPA 70E* to explore the requirements in the four categories of alerting techniques of 130.7(E) that were introduced in the list above. These and all related requirements must be referred to in their entirety for a full and complete understanding.

OVERHEAD LINES

Section 130.8 deals with overhead line work—specifically, requirements for work within the limited approach boundary or arc flash boundary of overhead lines. This section is divided into six categories. These and all requirements related to this work must be referred to in their entirety for a full and complete understanding. These six categories are:

- Uninsulated and Energized
- Determination of Insulation Rating
- Employer and Employee Responsibility; **see Figure 6-27**
- Deenergizing or Guarding; **see Figure 6-28**
- Approach Distances for Unqualified Persons
- Vehicular and Mechanical Equipment

Notice that the only mention of where the requirements of 130.8 apply is in the title—that is, Work Within the Limited Approach Boundary or Arc Flash Boundary of Overhead Lines. Review the definition of each of those boundaries in Article 100, their application in 130.4, and associated distances in Table 130.4(D)(a) for ac and Table 130.4(D)(b) for dc.

The note for superscript c in Table 130.4(D)(a) and the note for the superscript asterisk in Table 130.4(D)(b) advise, in part, that the limited approach boundary for an exposed movable conductor normally applies to overhead line conductors supported by poles. Recall

Figure 6-27. EMPLOYER AND EMPLOYEE RESPONSIBILITY

Figure 6-27. *The requirements for work within the limited approach boundary or arc flash boundary of overhead lines within the scope of* NFPA 70E *address topics including employer and employee responsibility. Courtesy of Service Electric Company.*

that the arc flash boundary is either the distance where the incident energy equals 1.2 cal/cm^2 per 130.5(B)(1) or determined by Table 130.7(C)(15)(A)(b) for ac systems or Table 130.7(C)(15)(B) for dc systems per 130.5(B)(2).

The requirements of 130.8 reference both qualified persons and unqualified persons. Recall that training requirements for both qualified and unqualified persons are located in 110.2(D) and that both are defined in Article 100.

Be sure to refer to *NFPA 70E* to explore the requirements in the six categories of 130.8 that were introduced here.

Recognizing that the 130.8 provisions—like all of Article 130—are not independent of the rest of the requirements of *NFPA 70E*, be sure to reference these and apply these and all applicable *NFPA 70E* rules and definitions, including those in Articles 90, 100, 110, 120, and 130.

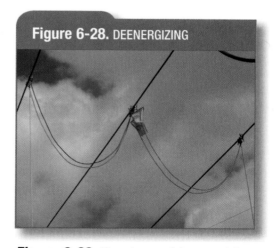

Figure 6-28. DEENERGIZING

Figure 6-28. *The deenergizing provisions in 130.8 require that arrangements be made with the person or organization that operates or controls the lines to deenergize them and visibly ground them at the point of work if the lines are to be deenergized. Courtesy of Service Electric Company.*

UNDERGROUND ELECTRICAL LINES AND EQUIPMENT

NFPA 70E includes provisions that introduce a number of concepts from OSHA's 29 CFR Part 1926 requirements for underground installations.

1926.651(b)
Underground installations.
1926.651(b)(1)
The estimated location of utility installations, such as sewer, telephone, fuel, electric, water lines, or any other underground installations that reasonably may be expected to be encountered during excavation work, shall be determined prior to opening an excavation.
1926.651(b)(2)
Utility companies or owners shall be contacted within established or customary local response times, advised of the proposed work, and asked to establish the location of the utility underground installations prior to the start of actual excavation. When utility companies or owners cannot respond to a request to locate underground utility installations within 24 hours (unless a longer period is required by state or local law), or cannot establish the exact location of these installations, the employer may proceed, provided the employer does so with caution, and provided detection equipment or other acceptable means to locate utility installations are used.
1926.651(b)(3)
When excavation operations approach the estimated location of underground installations, the exact location of the installations shall be determined by safe and acceptable means.
1926.651(b)(4)
While the excavation is open, underground installations shall be protected, supported or removed as necessary to safeguard employees.

NFPA 70E addresses this topic in Section 130.9 and requires, in part, that, before excavation starts, and where there exists a reasonable possibility of contacting electrical lines or equipment, the employer must take the necessary steps to contact the appropriate owners or authorities to identify and mark the location of the electrical lines or equipment. Review this requirement in its entirety including determining when appropriate safe work practices and PPE must be used.

CUTTING OR DRILLING

Section 130.10 is the last section on Article 130 and Chapter 1. It addresses cutting or drilling into equipment, floors, walls, or structural elements where a likelihood of contacting energized electrical lines or parts exists. Review the provisions of 130.10 in their entirety and consider how the risk assessment required to be performed by the employer will identify and mark locations, create an electrically safe work condition, and identify necessary safe work practices and PPE.

These are among tasks that electrical workers perform with some frequency. As always, consider all of the requirements of *NFPA 70E* that apply to this requirement. These may include 110.1(G), 110.4(B), 120.1, 130.2, 130.2(A), 130.4, 130.5, and a host of others.

Summary

This chapter covered requirements related to work involving electrical hazards through overview, abbreviated content, and excerpts. Article 130 needs to be referenced to look at these requirements in their entirety.

The Article 130 requirements are broken down into ten sections covering topics including, but not limited to, those related to demonstration of justification for energized work, the energized electrical work permit, a shock hazard analysis, an arc flash hazard analysis (including equipment labeling), personal protective equipment, and other protective equipment.

These requirements, like all *NFPA 70E* requirements, apply within the scope of the standard spelled out in Article 90. Remember to review and apply all pertinent definitions in Article 100, the general requirements of Article 110, and the requirements for establishing an electrically safe work condition in Article 120. In addition, review and apply any relevant *NFPA 70E* Chapter 2 and 3 requirements and any of the related information in the Informative Annexes.

Review Questions

1. Incident energy analysis is defined as a component of an arc flash risk assessment used to predict the incident energy of an arc flash for a(n) __?__ set of conditions.
   a. general
   b. nominal
   c. specified
   d. unspecified

2. An energized electrical work permit is only required when work is performed within the restricted approach boundary.
   a. true
   b. false

3. Appropriate safety-related work practices must be determined before a person is exposed to the electrical hazards involved by using both a shock risk assessment and an arc flash risk assessment.
   a. true
   b. false

4. One reason why a shock risk assessment is required is to determine the personal protective equipment necessary to __?__ the possibility of electric shock.
   a. eliminate
   b. maximize
   c. minimize
   d. remove

5. Under no circumstance is an unqualified person permitted to cross the __?__ approach boundary.
   a. arc flash
   b. limited
   c. restricted
   d. secured

6. The arc flash hazard risk assessment must take into consideration the design of the overcurrent protective device and its opening time, __?__ its condition of maintenance.
   a. eliminating
   b. excluding
   c. ignoring
   d. including

7. Which one of the following is not among the category of requirements related to "other precautions for personnel activities" in 130.6?
   a. Alertness, blind reaching, and confined or enclosed work spaces
   b. Conductive articles being worn, clear spaces, and doors and hinged panels
   c. Electrical hazard, arc blast, and cost benefit analysis
   d. Illumination, housekeeping duties, and anticipating failure

8. Shock protection must be worn within the __?__ approach boundary and arc flash protection must be worn within the __?__ boundary.
   a. limited, arc blast
   b. limited, arc flash
   c. restricted, limited
   d. restricted, arc flash

9. When using arc flash PPE categories to select personal protective equipment, it is important to note the parameters that must be met, such as minimum working distance, estimated short-circuit current, and __?__ fault clearing times, to determine the appropriate protective equipment.
   a. anticipated
   b. approximated
   c. maximum
   d. minimum

10. An arc flash PPE category __?__ be determined through an incident energy analysis.
   a. can
   b. cannot
   c. should
   d. should not

Bolted and Arcing Fault Current and Reading Time-Current Curves

Chapter Outline

- Short-Circuit Current
- OCPD Clearing Time

Chapter Objectives

1. Recognize two types of short-circuit (fault) current—bolted and arcing.
2. Understand the relationship between bolted and arcing fault current.
3. Understand how to determine fault-clearing times from OCPD time–current curves.

Chapter 7

Reference

1. *NFPA 70E®*, 2015 edition

Case Study

A 33-year-old male electrical engineer died from injuries sustained in an electrical fire while he was trying to get the serial number from an electrical panel. The victim received second- and third-degree burns to 90% of his body due to an explosion in an electrical panel. The victim lived for six days after the incident occurred.

The victim had come in for his final paycheck and was no longer employed by the company at the time of the incident. He went to investigate a problem concerning an electrical panel that had caused a power outage and an employee injury one hour earlier that morning. During the incident, coworkers heard screams and went to the location where the victim was located. No one witnessed the actual incident. When the coworkers arrived at the site, they found the victim on fire and screaming that he had not touched anything.

Interviews were conducted during the following week, including an interview with the employee who was injured in the earlier incident on the day of the fatality. The building owner refused to be interviewed and would not allow an on-site investigation to take place. The employer had been working at this location for approximately 10 weeks, and was in the process of moving when the incident occurred. It belonged to an industry that uses many locations for short periods of time and then moves on. The building had a prior history of electrical problems, according to both the Los Angeles Building and Safety office and the owner of a car dealership located downstairs below the incident site.

The Medical Examiner's report stated that the cause of death was sepsis due to multi-organ failure, secondary to massive thermal burns and inhalation injury.

Source: For details of this case, see California Case Report 92CA00201. Accessed October 23, 2012

For additional information, visit qr.njatcdb.org Item #1197.

INTRODUCTION

In this chapter, the concepts of bolted short-circuit (fault) current, arcing fault current, and overcurrent protective device (OCPD) opening time or clearing time are presented. In addition, the chapter provides a brief overview of how to read OCPD time–current curves.

This background information is necessary to prepare for the chapter "Methods to Accomplish the Arc Flash Risk Assessment." The depth of knowledge an individual must master from this chapter depends on the role the person has in electrical safe work practices related to 130.5 Arc-Flash Risk Assessment. This material is important for those workers who must determine arc-flash boundaries (AFB) (130.5(B)), arc-flash incident energies (130.5(C)(1)), or arc flash PPE categories (130.5(C)(2)).

When performing an arc-flash risk assessment to determine the AFB and arcing fault incident energy (for the appropriate PPE arc rating), it is important to note that the available short-circuit (fault) current and the OCPD clearing time are parameters that affect the results. For arcing fault incident energy calculations, the OCPD clearing time depends on the OCPD characteristics and the arcing fault current magnitude. The magnitude of the possible arcing fault current depends on the available short-circuit (bolted) current.

When performing an arc-flash risk assessment to determine the AFB and PPE category using the arc flash PPE categories method, the maximum short-circuit current available and the maximum fault clearing time for the type or family of OCPD must be within the parameter values specified in Table 130.7(C)(15)(A)(b); meeting these criteria is a condition of use of the table.

SHORT-CIRCUIT CURRENT

Overcurrent is defined as either an overload current or a short-circuit current, which often is referred to as a fault current.

Overload current is an excessive current relative to normal operating current, but one that is confined to the normal conductive path provided by the conductors, circuit components, and loads of the distribution system. Harmless overloads are routinely caused by temporary inrush or surge currents that occur when motors are started or transformers are energized. Such temporary overload currents are normal occurrences. In contrast, potentially harmful overloads can result from improperly designed or operated equipment, worn equipment, or too many loads on one circuit.

A sustained overload current results in overheating of conductors and other circuit components or adjacent materials and can cause deterioration of insulation, which may eventually result in severe damage, fires, and short circuits, if the overload current is not interrupted in sufficient time to prevent damage.

As the name implies, a short-circuit current is a current that flows outside the normal conducting path. One generally accepted definition of short circuit is when a phase or ungrounded conductor comes in contact with, or arcing current flows between, another phase conductor, neutral, or ground.

Whereas overloads are modest multiples of the normal current, a short circuit can be many hundreds or thousands of times larger than the normal operating current. The magnitude of available bolted short-circuit currents in commercial and industrial facilities varies dramatically, from roughly 1,000 amperes to more than 200,000 amperes.

Two categories of short-circuit (fault) currents are distinguished: bolted and arcing. In addition, a bolted or arcing short circuit can occur between various parts of a circuit, such as line-line (L-L), line-line-line (L-L-L), line-neutral (L-N), line-ground (L-G), or any combination of L, N, and G.

Bolted Short-Circuit (Fault) Currents

A bolted fault condition represents a "solid" (bolted or welded) connection of relatively low (or assumed "zero") impedance. Because of the low-impedance path, a 3-phase bolted fault condition is typically assumed to be the highest

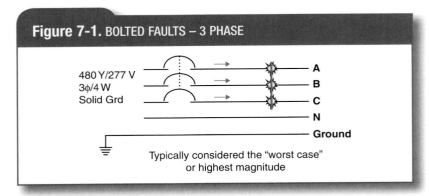

Figure 7-1. BOLTED FAULTS – 3 PHASE

Figure 7-1. A 3-phase bolted fault on a 3-phase, 4-wire solidly grounded system is typically considered the "worst case" or highest magnitude of fault current. *Courtesy of Eaton's Bussmann Business.*

magnitude fault current. The exceptions to this assumption are not covered in this chapter. **See Figure 7-1.** A line-to-line bolted fault current is 87% of the 3-phase bolted fault current. **See Figure 7-2.** Bolted line-to-ground fault currents can range from 25% to 125% of the 3-phase value, depending on the distance from the source or transformer.

The magnitude of a bolted short-circuit current is a function of the system voltage and the impedance for the electrical source to the point of the fault. In an electrical distribution system, the bolted short-circuit current varies depending on the point in the system at

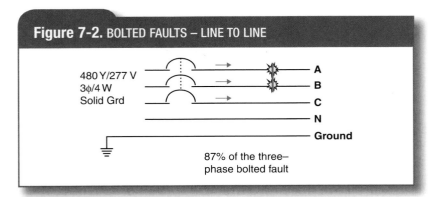

Figure 7-2. BOLTED FAULTS – LINE TO LINE

Figure 7-2. In a line-to-line fault on a 3-phase, 4-wire solidly grounded system, the current will be approximately 87% of the 3-phase bolted fault current. *Courtesy of Eaton's Bussmann Business.*

which the fault occurs. Calculating the bolted short-circuit current that would flow if a bolted fault occurred in a system is a well-established procedure that is commonly performed. The available (bolted) short-circuit current is needed to ensure that OCPD interrupting ratings (*NEC* 110.9, *NFPA 70E* 210.5, and OSHA §1910.303(b)(4)) and equipment short-circuit current ratings (*NEC* 110.10 and OSHA §1910.303(b)(5)) are greater than the available short-circuit current where they are applied. Also, performing coordination studies and ensuring the choice of OCPDs are selectively coordinated (*NEC* 700.28 and other sections) require knowing the available short-circuit currents. As will be established in this chapter, the arcing fault current magnitude for arc-flash risk assessment is directly related to the available short-circuit current.

A simple 3-phase electrical system is depicted as a one-line diagram. **See Figure 7-3.** The highest available 3-phase short-circuit current is typically found at the service transformer's secondary terminals and service equipment. The available 3-phase short-circuit currents at various points in the system are shown with "X" symbols, with the value for the symmetrical root mean square (rms) amperes denoted by I_{SC} (such as X_3 with 15,000 amperes available short-circuit current). In addition, the value for the 3-phase arcing fault current, expressed in rms amperes, at each point in the system is shown with "X" symbols (such as 9,270 amperes arcing fault current at X_3).

The chapter "Fundamentals of 3-Phase Bolted Fault Currents" covers available short-circuit current in more detail and provides a calculation method and examples.

Arcing Faults

An arcing fault condition does not have a solid bolted connection between conductors or buses; instead, an arcing current path through the air between conductive parts with an associated arc resistance is present. **See Figure 7-4.** Arcing faults can be 3-phase, phase to

Figure 7-3. ONE-LINE DIAGRAM OF A 3-PHASE SYSTEM

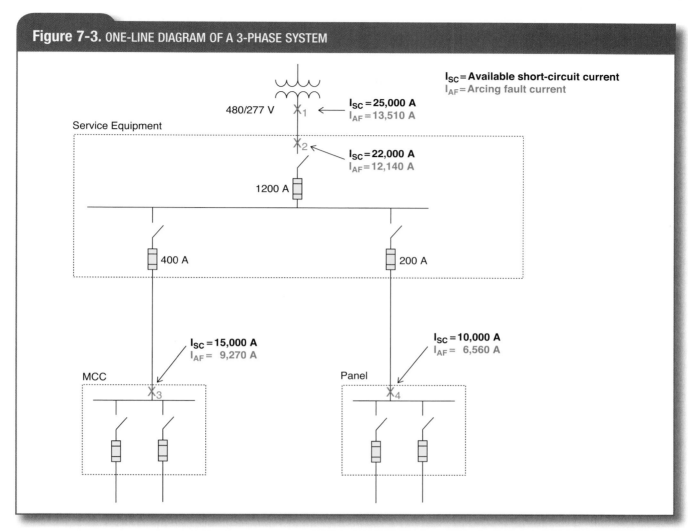

I_{SC} = **Available short-circuit current**
I_{AF} = **Arcing fault current**

480/277 V   1
I_{SC} = **25,000 A**
I_{AF} = **13,510 A**

Service Equipment

2
I_{SC} = **22,000 A**
I_{AF} = **12,140 A**

1200 A

400 A   200 A

I_{SC} = **15,000 A**
I_{AF} = **9,270 A**

I_{SC} = **10,000 A**
I_{AF} = **6,560 A**

MCC   Panel

3   4

Figure 7-3. This system is depicted as a simplified one-line diagram with available 3-phase short-circuit current and 3-phase arcing fault current examples noted at key points.

phase, phase to ground, or phase to neutral, or can include one or more phases with ground and/or neutral.

An arcing fault current is always less than the available bolted fault current value at a specific location due to the additional resistance of the arc. Its magnitude depends on many variables. In some cases, an arcing current is not sustainable, meaning that it will self-extinguish. A self-sustaining arcing fault, in contrast, may persist for a relatively long time (several cycles or longer). The ability of an arc to be initiated and sustained reflects many variables, including whether the

Figure 7-4. ARCING FAULT

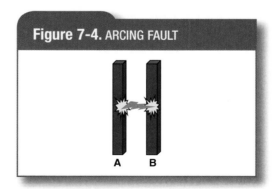

A   B

Figure 7-4. An arcing fault current is characterized by current flow via an arc through the air.

fault is on a single-phase or 3-phase circuit, the system voltage, the length of the arc (arc gap), the available bolted short-circuit current, and whether the arcing source is within an enclosure (for instance, switchgear cubicle) or in the open (for instance, utility distribution conductors on an outdoor pole). The size of the enclosure also makes a difference: The smaller the enclosure, the easier it is for the arc to continue. For example, an arc that self-extinguishes in a Size 5 motor starter may be self-sustaining in a Size 1 motor starter.

Generally, for low-voltage systems that experience high-thermal-intensity arcing fault events, if an arcing fault is initiated on a 1-phase circuit (L-L, L-N, or L-G on a 1-phase system) or an L-L, L-N, or L-G circuit (on a 3-phase system without the third phase being present), the arcing current may very well self-extinguish (that is, not sustain itself). By comparison, an arcing fault current that is initiated on a 3-phase circuit as a L-L arcing fault, L-N arcing fault, or L-G arcing fault can quickly (in less than one-thousandth of a second) escalate into a 3-phase arcing fault. **See Figure 7-5.** However, a 3-phase arcing fault current may or may not sustain itself based on the parameters of the electrical circuit and variables of the fault conditions, such as arc-gap spacing, the current's containment within an enclosure, the size of enclosure, the

available bolted short-circuit current, and other factors. For medium and high voltage systems, single phase arcing fault currents are more readily sustainable.

While calculating the 3-phase bolted short-circuit current available is rather straightforward and several methods and industry tools are available to assist with this process, calculating 3-phase arcing fault currents is not as straightforward and the results are not as certain. This ambiguity arises because of the nature of arcing faults as well as the many variables pertaining to the immediate surroundings that can affect an arcing fault event.

A direct relationship exists between the magnitude of the available bolted short-circuit current and the arcing fault current. For a given point in the electrical system, the arcing fault current is less than the available bolted short-circuit current. The greater the available short-circuit current, the greater the arcing fault current. However, the relationship between the magnitude of the available bolted short-circuit current and the magnitude of the arcing fault is not linear. In other words, the arcing fault current is not a fixed percentage of the available short-circuit current over the range of low to high available short-circuit currents.

Two different methods can be used to determine the arcing current. Both methods are based on first knowing the available short-circuit current (bolted short-circuit current). The first method is an older and for the most part out dated method, which is described in D.3 of *NFPA 70E* Informative Annex D. With this method two values of 3-phase arcing current are assumed (maximum and minimum) and then using the equations in D.3, the incident energy is calculated for each assumed arcing current value. The maximum arcing current value is assumed to be the same as the available short-circuit current (bolted) and the minimum arcing current value (for 480V systems) is assumed to be 38% of the available short-circuit current. The time duration for each scenario is determined using each of these arcing current values and reading the clearing time from the

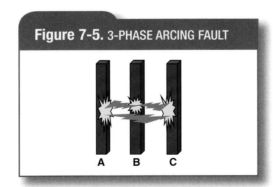

Figure 7-5. 3-PHASE ARCING FAULT

Figure 7-5. A 3-phase arcing fault has arcing current from phase A to phase B, from phase B to phase C, and from phase C to phase A (in addition, the arcing current can go from any of the phases to ground or neutral).

OCPD time-current curve. Years ago this method of calculation provided some information about the arcing current, but in today's environment its results may not be sufficient.

The second method, a more accurate calculation method, based on extensive arc-flash testing, is now available in the 2002 Institute of Electrical and Electronics Engineers (IEEE) 1584, Guide for Performing Arc-Flash Hazard Calculations. (The IEEE 1584 equations discussed in this chapter are from IEEE 1584 Guide. © 2002, by IEEE. All rights reserved.) The arcing current can be calculated by the formulas found in *NFPA 70E* Informative Annex D.4.2, taken from IEEE 1584 (note that the term "arc-in-a-box" refers to an arc that occurs within a confined space, such as when an arcing fault occurs within a motor control center bucket or in a panelboard).

The arcing fault currents for X_1 to X_4 (Figure 7-3) were determined by using the 2002 IEEE 1584 equation in *NFPA 70E* Informative Annex D, equation D.4.2(a) with the respective available short-circuit currents and other appropriate parameters. The variables in D.4.2(a) are important to understand. The constant K is selected based on arc in open air or arc in a box. G is the gap in mm which are provided in Table D.4.2. The gap represents the spacings between conductive parts of different phases for various types of electrical equipment.

Using the formula for systems less than 1 kilovolt, the 3-phase arcing current values can be calculated and compared to the available 3-phase short-circuit current. **See Figure 7-6.** This table is valid only for arcs-in-a-box, 480-volt systems, and applies to switchgear, motor control centers (MCCs), or panelboards. Note that the arcing current in this table varies from 43% to 83% of the 3-phase available short-circuit current value for switchgear at 480 volts or from

Figure 7-6. COMPARISON OF ARCING FAULT CURRENT AS A PERCENTAGE OF AVAILABLE SHORT-CIRCUIT CURRENT

480-Volt Switchgear (1.25" arc gap)			480-Volt MCC/Panelboard (1" arc gap)		
3-Phase Short-Circuit Current (kA)	3-Phase Arcing Current (kA)	Percentage Arcing Current to Short-Circuit Current	3-Phase Short-Circuit Current (kA)	3-Phase Arcing Current (kA)	Percentage Arcing Current to Short-Circuit Current
2	1.65	83	2	1.66	83
5	3.53	71	5	3.63	73
10	6.30	63	10	6.56	66
20	11.22	56	20	11.85	59
30	15.72	52	30	16.76	56
40	19.98	50	40	21.43	54
50	24.06	48	50	25.93	52
60	28.01	47	60	30.30	51
70	31.85	46	70	34.57	49
80	35.59	45	80	38.74	48
90	39.26	44	90	42.84	48
100	42.86	43	100	46.88	47

Figure 7-6. A comparison of the arcing fault current-in-a-box calculated using IEEE 1584 equations for various levels of available short-circuit currents for two different equipment types: 480-volt switchgear (1¼" conductor gap) and 480-volt MCC/panel (1" conductor gap). Review and compare the percentage 3-phase arcing current to 3-phase available short-circuit (bolted) current.

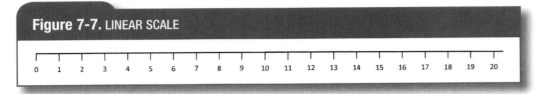

Figure 7-7. LINEAR SCALE

| | | | | | | | | | | | | | | | | | | | | |
|0|1|2|3|4|5|6|7|8|9|10|11|12|13|14|15|16|17|18|19|20|

Figure 7-7. *A linear scale example illustrates the equal distance between each incremental number.*

47% to 83% for MCC or panelboards at 480 volts. Also, note that the arcing current as a percentage of the 3-phase available short-circuit current (bolted) increases as the available short-circuit current magnitude decreases.

OCPD CLEARING TIME

The OCPD clearing time is an important factor in the resultant level of incident energy calculated or the determination of arc flash PPE category. Determining an OCPD clearing time typically involves reading OCPD time–current curves.

Reading OCPD Time–Current Curves

This section describes how OCPDs respond to various levels of overcurrent and is an introduction on how to read OCPD time–current curves. It represents a simplified introduction to the topic. It does not cover how to plot OCPD time–current curves.

It is important to understand that the concept of using fuse and circuit breaker time–current curves to determine the opening time is applicable to any type of overcurrent. For overcurrent protection, fuses and circuit breakers sense and respond to the magnitude of the actual current that flows through the OCPD, whether the current is a bolted or an arcing fault current.

Log – Log Graphical Representation of OCPD Characteristics

OCPD are commonly represented graphically with the horizontal axis for current and the vertical axis for time. Typically a logarithmic scale is used for both the current and time axis rather than linear scales. To read values plotted on a log – log graph merely requires understanding this type of representation and how it differs from linear – linear graphs.

In our educational system and life, linear scale is a familiar graphical representation. **See Figure 7-7.** A linear scale has equal distance between incremental numbers such as 0, 1, 2, 3, 4, 5, 6, 7, 8, 9, 10 and so on.

Logarithmic scale is a better method to use when the data values to be represented cover a wide range of values, such as an OCPD time current curve. For a base 10 logarithmic scale, which is used for time current curves, the exponents of 10 such as 1, 10, 100, and 1,000 are equally spaced. However, for values between the exponents of 10 numbers, the distances between numbers such as 1, 2,

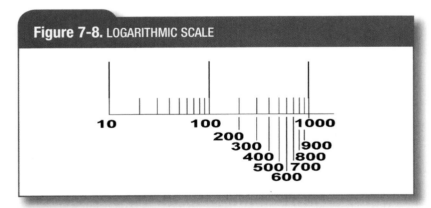

Figure 7-8. LOGARITHMIC SCALE

Figure 7-8. *A logarithmic scale example illustrates equal distance between the values of 10, 100, and 1000, however, varying distances between the intermediate values such as 100 to 200 compared to 900 to 1000.*

3, 4, 5, 6, 7, 8, and 9 vary on the graph. **See Figure 7-8.** Notice the distance between 10 and 100 is equal to the distance from 100 to 1000. However, the distance from 100 to 200 is not equal to the distance from 200 to 300.

It is impractical to represent OCPD time-current curves by a linear – linear scale representation because it would result in a range of values too large. OCPD time-current curves are represented on log – log graphs. **See Figure 7-9.** This is a time-current curve for a 100-ampere fuse. The horizontal axis represents the current in amperes and the vertical axis represents the time in seconds. Notice the current axis values of 10, 100, 1,000, 10,000, and 100,000 amperes are equal distance from each other. Then review the distance between 100, 200, 300, 400, 500, 600, 700, 800, 900, and 1000 amperes; these distances progressively are less as the magnitude increases. The vertical axis represents time in seconds starting at 0.01 second up to 1000 seconds. The distance is the same between 0.01,

Figure 7-9. TIME-CURRENT CURVE

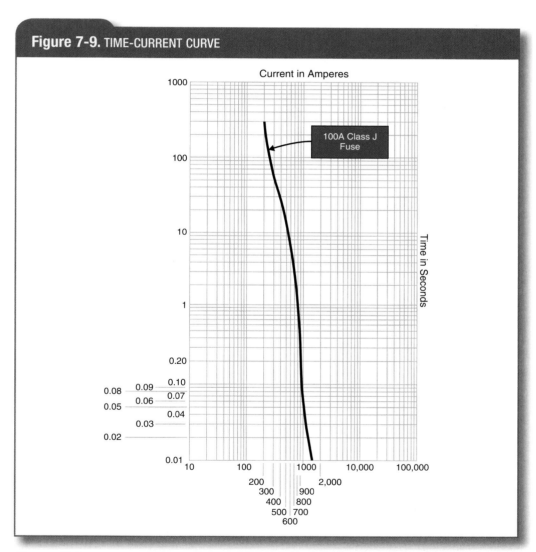

Figure 7-9. *The time–current curve characteristic for the 100-A fuse (100-A Class J fuse) has logarithmic scale for the current on the horizontal axis and for the time on the vertical axis.*

Class J fuses are available in 30, 60, 100, 200, 400, and 600 ampere case sizes.

0.1, 1, 10, 100, and 1000 seconds. The distances between 0.01, 0.02, 0.03, 0.04, 0.05, 0.06, 0.07, 0.08, 0.09, and 0.1 progressively are less as the time increases.

Fuse Clearing Times

A circuit supplies a load protected by a 100-ampere fuse under different overcurrent scenarios, designated as A to E. **See Figure 7-10.** Scenario A: A 100-A fuse in an electrical enclosure should carry an 80-A load continuously without opening. Scenario B: A 300-A overload (300% of the 100-A fuse ampere rating) is interrupted by the 100-A fuse in approximately 50 seconds. Scenario C: A 500-A overload (500% of the 100-A fuse ampere rating) is interrupted by the 100-A fuse in approximately 12 seconds. Scenario D: A 1,500-A fault current (15 times the 100-A fuse rating) is cleared by the 100-A fuse in approximately 0.01 second (approximately one-half cycle). Scenario E: A 20,000-A fault current is cleared by the 100-A fuse in less than one-half cycle (the fuse is in its current-limiting range).

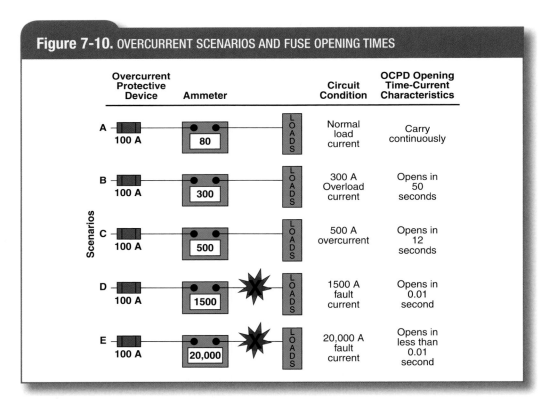

Figure 7-10. Different overcurrent scenarios and opening times of a 100-A fuse are demonstrated on a circuit diagram.

An ammeter reads the circuit current for each circuit condition and the 100-A fuse has a clearing time associated with each magnitude of overcurrent.

One characteristic of fuse technology is that the higher the percentage overcurrent, the faster the fuse opens. This relationship is referred to as an inverse time–current characteristic: The higher the current level during a fault, the faster the reaction time or interrupting time of the overcurrent device.

The opening time of the 100-A fuse is determined from the time–current characteristic curve for this specific fuse.

For a given overcurrent value, the fuse time–current curve can be used to determine the fuse clearing (interrupting or opening) time. **See Figure 7-11.** Note that in the example with an 80-A continuous current, the fuse will carry this current indefinitely. The 100-A fuse protects the circuits for the scenarios designated as A to E; note the different overcurrent

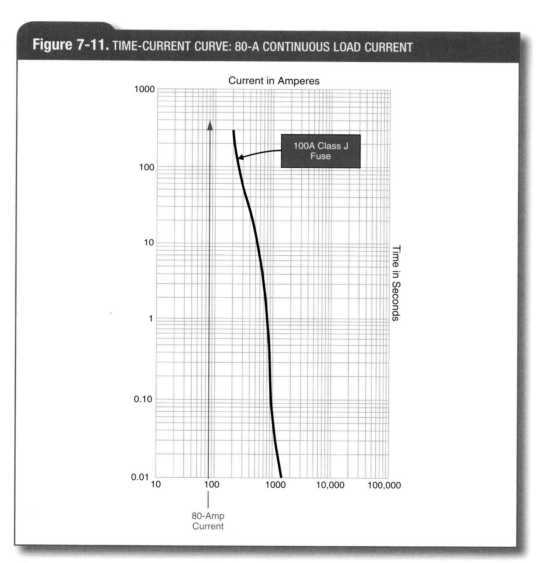

Figure 7-11. TIME-CURRENT CURVE: 80-A CONTINUOUS LOAD CURRENT

Figure 7-11. The 100-A fuse protecting the circuit for scenario A should carry an 80-A continuous-load current indefinitely.

values and clearing times. **See Figures 7-12, 7-13, 7-14, and 7-15.**

The speed of response of an OCPD when interrupting an overcurrent greater than the OCPD's ampere rating can vary depending on the magnitude of overcurrent. To explore that concept, begin with the principle that OCPDs are intended to continuously carry the load current, which is typically below the ampere rating of the OCPD. If the overcurrent is a light overload, slightly higher than the OCPD's ampere rating, it may be permissible to allow the current to flow for many minutes. Some circuit components—such as motors, primary windings of transformers, and capacitors—have a harmless high starting or energizing inrush current that can be many times greater than the normal full-load current. The OCPDs on these circuits must permit intentional short-duration overload currents for a period of time without opening. If the overcurrent is a faulted circuit, rapid OCPD

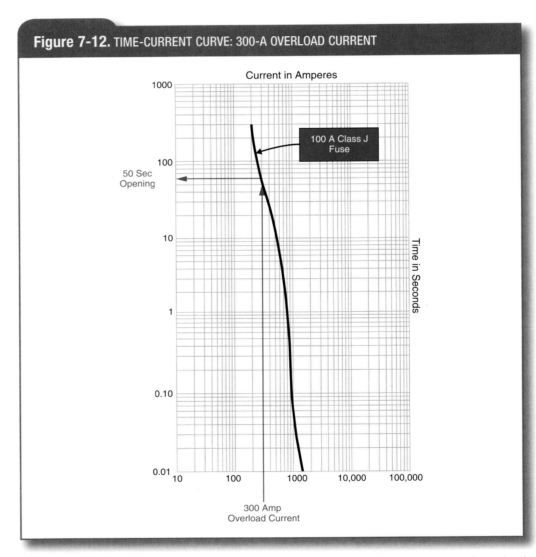

Figure 7-12. TIME-CURRENT CURVE: 300-A OVERLOAD CURRENT

Figure 7-12. The 100-A fuse protecting the circuit for scenario B clears the 300-A overload current in approximately 50 seconds.

response is desired to minimize circuit component or equipment damage.

Fuse time–current curves can vary based on the manufacturer and the type of fuses. For example, non-time-delay, time-delay, dual-element time-delay, and high-speed fuses are each associated with unique time–current curves. Nevertheless, all fuse curves demonstrate an inverse time relationship: The greater the overcurrent magnitude, the faster the clearing time.

Current-Limiting Fuse Clearing Times

The previous section on reading time–current curves provided basic information on how to determine fuse clearing times. Current-limiting fuses do not have the adjustability of overcurrent protection settings, which are found in some molded-case circuit breakers and low-voltage power circuit breakers. In most applications, there are choices in the fuse type that could be used. The choice of

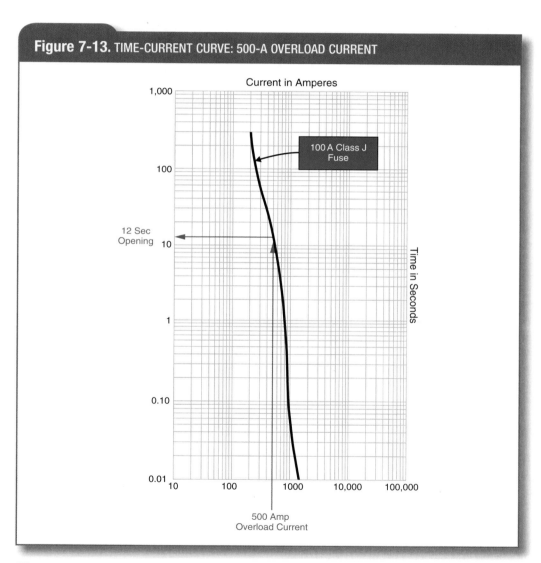

Figure 7-13. TIME-CURRENT CURVE: 500-A OVERLOAD CURRENT

Figure 7-13. The 100-A fuse protecting the circuit for scenario C clears the 500-A overload current in approximately 12 seconds.

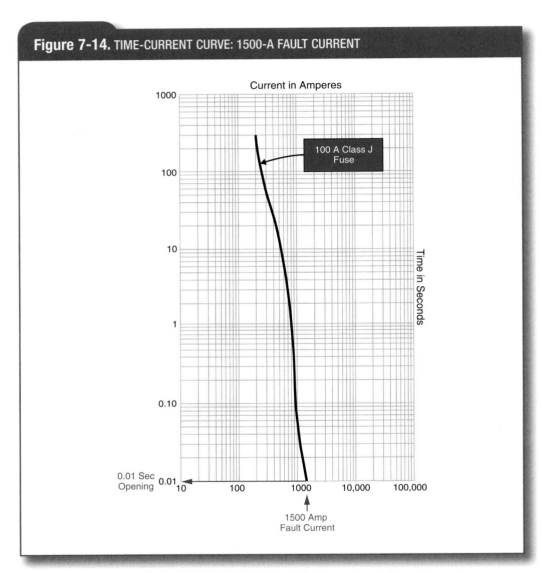

Figure 7-14. TIME-CURRENT CURVE: 1500-A FAULT CURRENT

Figure 7-14. The 100-A fuse protecting the circuit for scenario D interrupts the 1,500-A fault current in approximately 0.01 second (one-half cycle).

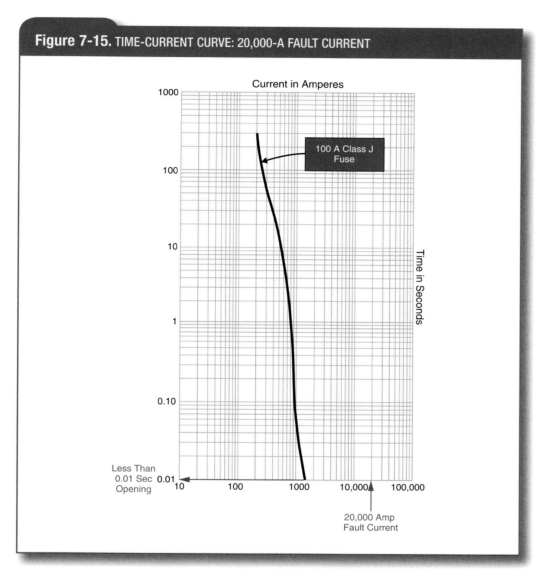

Figure 7-15. TIME-CURRENT CURVE: 20,000-A FAULT CURRENT

***Figure 7-15**. The clearing time for a 100-A fuse protecting the circuit for scenario E with a 20,000-A fault current cannot be determined by reading the time–current curve because the fuse is in its current-limiting range and clears in less than 0.01 second (less than one-half cycle).*

fuse type/characteristic may affect the arc-flash hazard results. For some types of loads, such as a motor branch circuit or transformer primary protection, different fuse types need to be sized differently.

Consider the time–current curves for four types of 400-A current-limiting fuses: Class RK5 dual-element time-delay, Class RK1 dual-element time-delay, Class J time-delay, and Class J non-time-delay. **See Figure 7-16.** All three time-delay fuse types have similar time-delay characteristics for times greater than 1 second. For times less than approximately 0.2 second, the 400-A Class RK5 fuse has a longer clearing time than the other fuses. For times less than 0.2 second, the

Class J time-delay fuse has the most advantageous time–current characteristic of all the time-delay fuses. The Class J non-time-delay fuse provides significantly faster clearing times for low-level overcurrents compared to the other fuses. However, for circuits with normal inrush currents, such as motor branch circuits and transformer primary protection, Class J non-time-delay fuses typically must have a larger ampere rating than Class J time-delay fuses.

One significant advantage of current-limiting fuses compared to standard molded-case circuit breakers and low-voltage power circuit breakers is that these fuses can provide reduced clearing

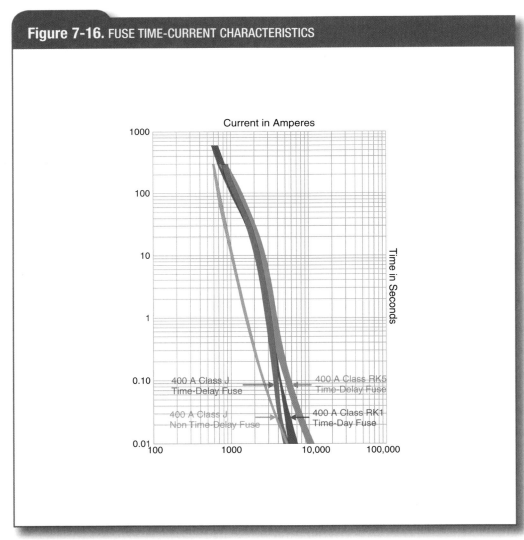

Figure 7-16. FUSE TIME-CURRENT CHARACTERISTICS

Figure 7-16. *Four types of 400-A current-limiting fuses may result in four different clearing times for some levels of overcurrent current.*

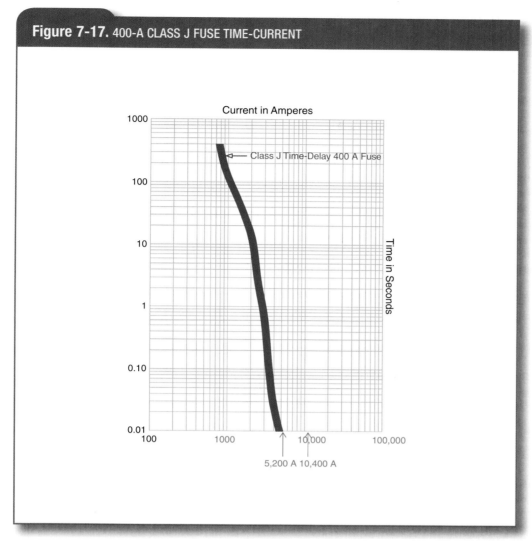

Figure 7-17. 400-A CLASS J FUSE TIME-CURRENT

Figure 7-17. This 400-A Class J fuse clears in one-half cycle or less for fault currents greater than 5,200 A. For a fault current of 10,400 A, it clears in approximately one-fourth cycle (0.004 second).

times when the arcing current is within the fuse's current-limiting range. The current-limiting range of a fuse is approximately where the current exceeds the current value at which the clearing time of the fuse is less than 0.01 second. This point is typically approximately 15 or less times the ampere rating of the Class J, RK1, CF, and T current-limiting fuses, but can be verified by examining the fuse time–current curve.

Current-limiting fuses are not always current-limiting, however: Similar to circuit breakers, they have an overload region in which their clearing time can be several seconds or minutes. When the arcing current is less than the current-limiting range of a fuse, extended clearing times and increased incident energy can occur. If the fault is within the current-limiting range, however, clearing times of less than 0.01 second (assumed to be one-half cycle or 0.008 second) occur. **See Figure 7-17.** Based on the testing conducted for IEEE 1584, the clearing time for current-limiting fuses can be assumed to be 0.004 second if the arcing current value is more than twice the current at 0.01 second. Note that a specific manufacturer's Class J 400-A fuse enters its current-limiting range at approximately 5,200 A (where the total clearing curve crosses 0.01 second); at 10,400 A, it is assumed to clear in 0.004 second or less. In addition, when a fuse operates in its current-limiting range, the let-through current is reduced from what is available. The combination of decreased clearing times and reduced arcing current can greatly reduce the incident energy when

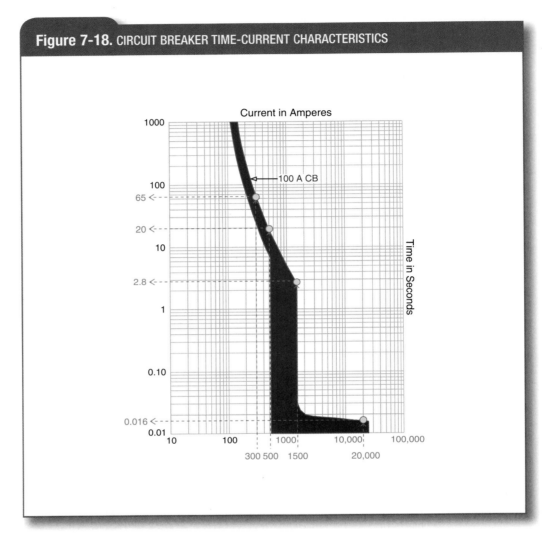

Figure 7-18. CIRCUIT BREAKER TIME-CURRENT CHARACTERISTICS

Figure 7-18. *Clearing times for various overcurrents for a 100-A thermal magnetic molded-case circuit breaker with a fixed instantaneous trip are shown. The dotted lines and yellow dots assist in determining the clearing times for various overcurrent values.*

Figure 7-19. 100-A CIRCUIT BREAKER

Overcurrent Magnitude	Clearing Time
300 A	65 seconds
500 A	20 seconds
1,500 A	2.8 seconds
20,000 A	0.016 second

Figure 7-19. *The 100-A circuit breaker results in these clearing times for the various overcurrent magnitudes.*

the arcing current is within the current-limiting range of a fuse.

Circuit Breaker Clearing Times

Circuit breaker time–current characteristic curves have a different shape than fuse time–current characteristic curves. Nevertheless, in most situations, similar methods can be used for reading circuit breaker and fuse time–current curves. **See Figures 7-18** and **7-19.** For the purposes of clearing time for arc-flash risk assessment, the greatest opening time for a given current should always be used; that is, use the line to the right and top of the curve. This is the maximum clearing time curve for a circuit breaker.

Several types of circuit breakers are available, and each may have time–current curve characteristics that differ in the shape or total clearing times for certain overcurrent values. To illustrate this point, the clearing time for a 20,000-A fault is shown for three different types of 600-A circuit breakers. Note the differences in the opening time. Given these variations, it is important to know which type of circuit breaker is being analyzed as well as which settings are in use.

Interpreting the time–current curve for a typical 600-A thermal-magnetic (TM) molded-case circuit breaker, the 20,000-A fault intersects the circuit breaker maximum clearing time at 0.024 second. **See Figure 7-20.** This circuit breaker has an adjustable instantaneous trip (IT) setting and in this case is set at 10 times the circuit breaker 600-A current rating—that

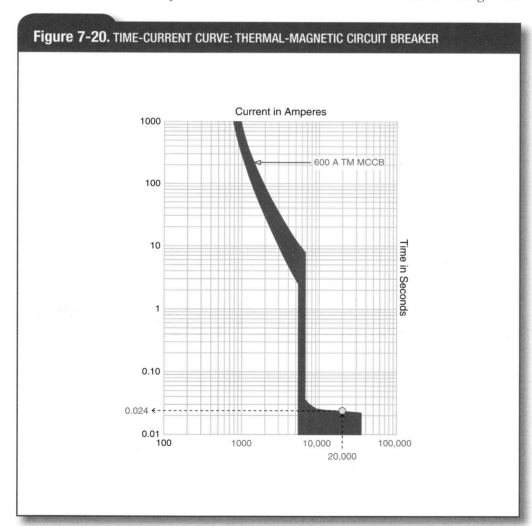

Figure 7-20. TIME-CURRENT CURVE: THERMAL-MAGNETIC CIRCUIT BREAKER

Figure 7-20. This 600-A thermal-magnetic molded-case circuit breaker clears a 20,000-A fault in 0.024 second.

is, at 6,000 A. A ± tolerance is associated with this setting, and for this particular case, the threshold of where the instantaneous trip should start to operate when set at 10 times the circuit breaker ampere rating is between 5,400 A and 6,600 A.

For molded-case circuit breakers with an adjustable instantaneous trip, the IT setting can typically be set as low as 3 to 5 times the circuit breaker ampere rating, but generally no higher than 8 to 12 times this rating. In addition, in the past decade, newer-style molded-case circuit breakers have become available with IT settings as high as 20 or more times the circuit breaker ampere rating.

The electronic-sensing unit of the 600-A electronic-sensing molded-case circuit breaker can offer increased accuracy and more curve shaping (adjustments). **See Figure 7-21.** Interpreting the time–current curve of a typical 600-A electronic sensing molded-case circuit breaker, the 20,000-A fault intersects the circuit breaker maximum clearing time at 0.04 second.

Electronic-sensing molded-case circuit breakers may have three electronic sensing regions: long-time (LT), short-time delay (STD), and instantaneous trip (IT). The "long-time" (LT) setting determines the ampere rating of the circuit breaker,

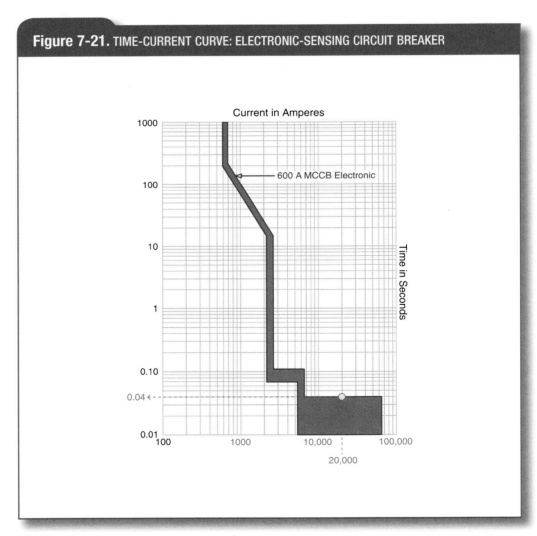

Figure 7-21. TIME-CURRENT CURVE: ELECTRONIC-SENSING CIRCUIT BREAKER

Figure 7-21. This 600-A electronic-sensing molded-case circuit breaker can clear a 20,000-A fault in 0.04 second.

whereas the "long-time delay" (LTD) setting determines the overload region performance. The short-time delay is a sensing region between the long-time region and the instantaneous region; it provides a means to not open the circuit breaker instantaneously for some levels of fault current, yet to open it faster than with the long-time delay. The short-time delay for electronic-sensing molded-case circuit breakers is typically available at 0.1 second (6 cycles), 0.2 second (12 cycles), or 0.3 second (18 cycles). In this case, the STD is set at 0.1 second.

The instantaneous trip for this electronic-sensing molded-case circuit breaker is set at 10 times the circuit breaker's 600-A current rating—that is, at 6,000 A. Note that with an electronic-sensing molded-case circuit breaker, the maximum IT setting cannot typically be more than 8 to 10 times the ampere rating of the circuit breaker. The 20,000-A fault current is in the instantaneous region and the clearing time is 0.04 second—slower than the clearing time (0.024 second) for the thermal-magnetic molded-case circuit breaker example. In some instances, electronic-sensing circuit breakers might open faster than a thermal-magnetic circuit breaker, but this behavior depends on the specific circuit breaker type and characteristics.

Many of the low-voltage power circuit breakers (LVPCB) are similar to electronic trip molded-case circuit breakers with regard to the adjustability of their overcurrent protection settings. The difference is that these devices are larger in size than molded-case circuit breakers, offer draw-out capabilities (for increased maintenance capabilities), and can be provided with or without an instantaneous trip. When the instantaneous trip is omitted (to increase their selective coordination capabilities), LVPCBs utilize a short-time delay. **See Figure 7-22.** This is the time-current characteristic curve for an older style LVPCB; modern LVPCBs use electronic sensing and have "tighter" tolerances and adjustability to "shape" the curve. In this case, the 20,000-A fault is cleared by the short-time delay in 0.5 second (30 cycles).

Electronic-sensing CBs can contain a human interface to the overcurrent adjustment trip unit. Courtesy of Eaton.

If equipped without an IT, the low-voltage power circuit breaker is set to "hold off opening" faults for a period of time equal to the short-time delay setting. The STD setting can range from 0.1 second (6 cycles) to 0.5 second (30 cycles). Its purpose is to allow the downstream OCPD to clear faults before the LVPCB trips, thereby avoiding unnecessary power loss to other loads. This increase of clearing time by using a short-time delay can result in an increase in arc-flash incident energy compared to the thermal-magnetic and electronic-sensing molded-case circuit breakers. As a consequence, for installations complying to the 2011 *NEC* or later edition, technology, such as zone-selective interlocking, arc-reduction maintenance switches, and arc-flash relays, is required when STD settings are used (without instantaneous trip) on low-voltage power circuit breakers (reference *NEC* 240.87). The chapter *Electrical System Design and Upgrade Considerations* provides information on these technology methods to mitigate the arc flash hazard.

Circuit breaker time–current characteristics may be either adjustable or nonadjustable. For instance, many 150-A or less rated molded-case circuit breakers used in branch panels do not have settings that can be adjusted; the instantaneous trip

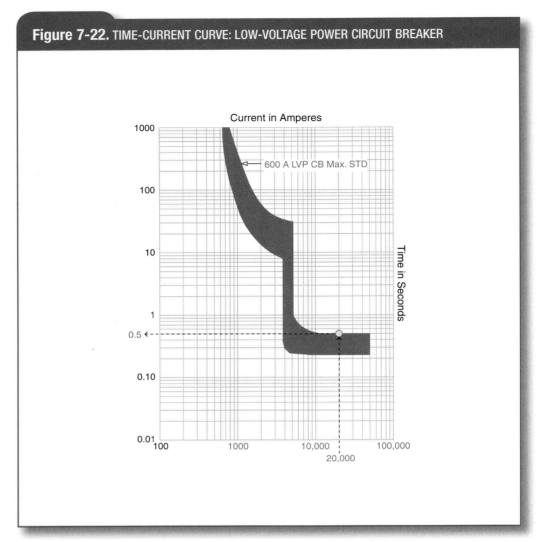

Figure 7-22. TIME-CURRENT CURVE: LOW-VOLTAGE POWER CIRCUIT BREAKER

Figure 7-22. This 600-A low-voltage power circuit breaker with a maximum short-time delay setting can clear a 20,000-A fault in 0.5 second. This demonstrates the time–current characteristics for a low-voltage power circuit breaker with a long-time region and a short-time delay region (no instantaneous trip).

settings are fixed. For circuit breakers with adjustable settings, the most common adjustments that can affect arc-flash hazard incident energy are changes made to the instantaneous trip settings and to the short-time delay settings. **See Figure 7-23.** A 400-A circuit breaker with an adjustable instantaneous trip is shown. Two different instantaneous trip settings are illustrated: 5 times (which is 2,000 A with a red tolerance band) and 10 times (which is 4,000 A with a blue tolerance band). An instantaneous trip setting can

influence the clearing time for low-level faults. In this case, for a 3,000-A fault, if the IT is set at 5 times the circuit breaker ampere rating, the circuit breaker clears in 0.024 second; by comparison, if the IT is set at 10 times this rating, the circuit breaker clears in 35 seconds.

Similar to current-limiting fuses, one significant advantage of current-limiting circuit breakers compared to standard molded-case circuit breakers and low-voltage power circuit breakers is that these circuit breakers can provide re-

Figure 7-23. TIME-CURRENT CURVE: 400-A CIRCUIT BREAKER: ADJUSTABLE INSTANTANEOUS TRIP

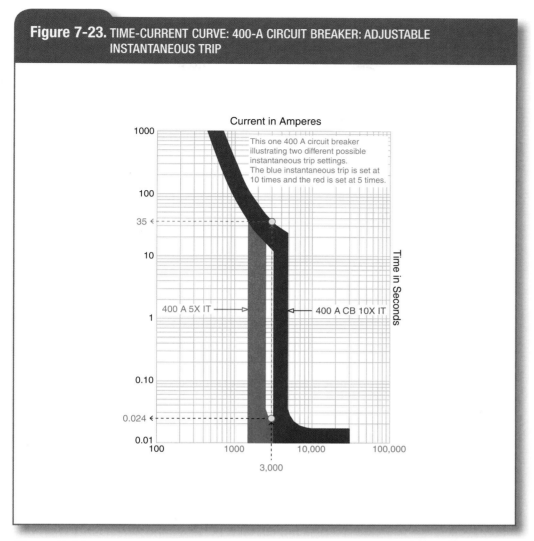

Figure 7-23. This 400-A molded-case circuit breaker clears a 3,000-A fault in 0.024 second when the instantaneous trip is set at 5 times the circuit breaker ampere rating and in 35 seconds when the instantaneous trip is set at 10 times this rating.

duced clearing times when the arcing current is within the circuit breaker's current-limiting range.

Generalized OCPD Clearing Times

As the previous sections have illustrated, fault clearing time can vary based on OCPD type, ampere rating, and settings as well as the magnitude of fault current. In determining the fault clearing time, it is recommended to use the time–current curve for the specific manufacturer, type, ampere rating, and setting of the OCPD.

Molded case circuit breakers come in various frame sizes.

Figure 7-24. GENERALIZED CLEARING TIMES

Device Type (600 V or less)	Clearing Time (seconds)[†]
Current-limiting fuse	0.004 to 0.008*
Molded-case circuit breaker: • Instantaneous trip • Short-time delay	0.025** Setting of short-time delay***
Insulated-case circuit breaker: • Instantaneous trip • Short-time delay	0.05** Setting of short-time delay***
Low-voltage power circuit breaker (integral trip) • Instantaneous trip • Short-time delay	0.05** Setting of short-time delay***
Current-limiting molded-case circuit breaker	0.008 or less*

[†]These are approximate clearing times for short-circuit currents within the current-limiting range of a fuse or a current-limiting circuit breaker or within the instantaneous region or short-time delay region of a circuit breaker.

*These are approximate clearing times for current-limiting fuses and current-limiting circuit breakers when the fault current is in their current-limiting range.

**These circuit breaker instantaneous trip times are based on Table 1 in 2002 IEEE 1584.

***Molded-case and insulated-case circuit breakers with a short-time delay setting typically also include an instantaneous trip override that operates above a certain fault current level.

Note: Lower current values may cause the overcurrent device to operate more slowly. Arc-flash energy may actually be highest at lower levels of available short-circuit current. This requires that arc-flash energy calculations be completed for the range of sustainable arcing currents. This is also noted in NFPA 70E 130.5 IN No. 2.

Figure 7-24. Generalized clearing times depend on the OCPD type.

See Figure 7-24. This table provides general OCPD clearing times. Typical short-time delay settings could be 6, 12, 18, 24, or 30 cycles.

OCPD Condition of Maintenance's Effects on Clearing Time

As noted in *NFPA 70E* 130.5(3), the arc-flash risk assessment must take into consideration the design of the OCPD and its clearing time, including its condition of maintenance. 130.5 Informative Note No. 1 provides the rationale. The design and condition of maintenance of OCPDs may directly impact the clearing time of OCPDs, which in turn affects the incident energy. Poorly maintained OCPDs may take longer to clear, or may not clear at all, resulting in higher actual arc-flash incident energies than those expected from an arc-flash risk assessment based on an OCPD operating as originally specified. If the design of an OCPD requires periodic testing and maintenance as well as inspection and testing after fault interruption (*NFPA 70E* 225.3) to ensure proper operation, inspection, maintenance, and testing must be performed and documented (*NFPA 70E* 205.4). If an OCPD's condition of maintenance cannot be assured to be acceptable, it may not be suitable to rely on the results of an arc-flash risk assessment. For more information on this topic, see the chapter "Maintenance Considerations and OCPD Work Practices."

Summary

An understanding of available short-circuit (fault) current, arcing current, OCPD time–current curves, and OCPD clearing times provides the basis for properly applying 130.5 Arc-Flash Risk Assessment. Different OCPD types can have differently shaped time–current characteristic curves, and some OCPDs have adjustable settings that can alter the shape of their time–current characteristic curve. The potential arc-flash incident energy determined from an arc-flash risk assessment is directly related to the OCPD type and characteristics.

Review Questions

1. Four different types of 400A fuses or four different type circuit breakers will always result in the same clearing time for a specific fault current.
   a. True
   b. False

2. A bolted fault condition represents a(n) __?__ connection of relatively low (or assumed "zero") impedance at the point of the fault.
   a. arcing
   b. dangling
   c. solid
   d. weak

3. For a given point in the electrical system, the greater the available short-circuit current, the __?__ the arcing fault current.
   a. faster
   b. greater
   c. less
   d. slower

4. Arc-flash incident energy varies by the fault current __?__.
   a. magnitude and duration
   b. instantaneous peak
   c. region
   d. vector analysis

5. Fuses and circuit breakers sense and __?__ to actual current magnitude, whether the current of a given magnitude is a bolted or an arcing fault current.
   a. close
   b. feel
   c. hesitate
   d. respond

6. The speed of response of an OCPD when interrupting an overcurrent greater than the OCPD's __?__ rating can vary depending on the magnitude of overcurrent.
   a. ampere
   b. power
   c. voltage
   d. wattage

7. The clearing time of OCPDs is a(n) __?__ factor in the severity of an arc-flash incident.
   a. important
   b. incidental
   c. insignificant
   d. minor

8. The arc-flash risk assessment takes into consideration the design of the OCPD and its clearing time, __?__ its condition of maintenance.
   a. concerning
   b. eliminating
   c. excluding
   d. including

9. One significant advantage of current-limiting circuit breakers or current-limiting fuses compared to standard molded-case circuit breakers and low-voltage power circuit breakers is that these circuit breakers and fuses can provide reduced clearing times when the arcing current is within their current-limiting range.
   a. True
   b. False

10. For a given point in a 480V system, the arcing fault current can be calculated using an IEEE 1584 equation, if the available fault current is known as well as some other parameters such as the type of electrical equipment (arc gap spacing).
    a. True
    b. False

Fundamentals of 3-Phase Bolted Fault Currents

Chapter Outline

- Short-Circuit Calculation Requirements
- Short-Circuit Calculation Basics
- Effect of Short-Circuit Current on Arc-Flash Risk Assessments
- Effect of System Changes on Arc-Flash Risk Assessments
- Procedures and Methods
- Point-to-Point Method

Chapter Objectives

1. Understand why short-circuit studies are required and when they are needed.
2. Recognize the effect of short-circuit current on arc-flash hazards.
3. Understand the procedures and calculation methods used to perform a 3-phase short-circuit current study.

Chapter 8

References

1. *National Electrical Code®* (*NEC®*), 2015 Edition
2. *NFPA 70E®*, 2015 edition
3. Bussmann SPD (Selecting Protective Devices) publication, section on Calculating Short-Circuit Currents
4. Appendix

Case Study

A customer needs to have a 200-ampere, 480-volt, 3-phase panel installed to feed a new addition. The new panel will be fed from an existing 600-A panel that has a 10,000-A short-circuit current rating (SCCR) and circuit breakers with 10,000-A interrupting ratings (IR). A few years ago, the existing panel was installed, presumably compliant with the *NEC*. The customer is asking contractors to quote a price for the required electrical equipment and installation.

Is it as simple as buying and installing a 3-phase, 480-V, 200-A, 10,000-A SCCR panel (with 10,000-A interrupting rated breakers) and installing it with the feeder from the existing 600-A panel (including installing a new 200-A circuit breaker in the existing panel)? Or is there more to this task?

What are some of the things that need to be considered? Considerations include the need to perform a short-circuit and arc flash study to verify compliance of the existing installation, select the proper new electrical equipment, and select shock and arc flash protection to install the new system. The new electrical equipment must be selected and installed in compliance with *NEC* requirements, including 110.9 and 110.10. Existing electrical equipment will require an examination for *NEC* compliance and must be replaced to comply with OSHA 1910.302(b)(1), 1910.303(b)(4), and 1910.303(b)(5) should analysis results indicate that the available fault current is now higher than the interrupting and short-circuit current rating of the existing equipment. Additional considerations may include electrical hazard analysis and work practices, as well as selection of personal protective equipment (PPE).

An analysis was performed that determined the existing installation has an available fault current greater than the existing panel SCCR and IRs for the circuit breakers. This situation may occur when either a lower impedance or larger size utility transformer is installed to replace an existing transformer.

Uncovering safety hazards of inadequate SCCR- and IR-rated existing equipment is unwelcome news to the facility owner but important as part of the effort to protect persons and property. Now the scope of work includes more than scheduling a shutdown so that the new 3-pole 200-A breaker can be installed in the existing panel. Not only is a panel with a more robust SCCR and IRs required than originally anticipated, but a decision now needs to be made by the customer on what to do with the existing equipment.

In this scenario, the situation was discovered before inadequately rated equipment became a factor. All too often, however, it is only after an incident occurs that the extent of the hazard is realized. Recognition, acceptance, and application of *NEC*, *NFPA 70E*, and OSHA requirements are helping to improve the safety levels of today's new and existing electrical systems.

All parties play an important role in ensuring a safe and reliable electrical installation, both today and in the future. Data collection is an essential part of analysis and must not be overlooked. The calculation and software options available today make it easier than ever to specify, install, and maintain Code- and OSHA-compliant installations.

Source: Steve Abbott, Stark Safety Consultants

For additional information, visit qr.njatcdb.org Item #1630.

INTRODUCTION

The available short-circuit current is identified through a short-circuit study. The purpose of a short-circuit current study is to determine the available short-circuit current (bolted) at one or multiple points in an electrical system so as to ensure safety compliance with OSHA, *NEC*, and *NFPA 70E* requirements, as well as for other electrical system analysis purposes. For arc flash risk assessment, typically the available short-circuit current is a necessary value in the process to determine the arc flash boundary (130.5(B)) and the arc flash PPE (130.5(C)), whether using the arc flash PPE categories method (table method) or incident energy analysis method.

This chapter provides some insight into the variables that affect the available short-circuit current magnitude at various points in an AC electrical system. These variables, such as switching schemes or magnitude of motor current contribution at time of fault, result in various scenarios. A thorough arc flash risk assessment includes evaluating all the potential scenarios for a particle system and generally choosing the results that result in the most conservative arc flash boundary and incident energy or arc flash PPE categories.

In addition, this chapter provides the procedures, including equations and tables, for performing simple short-circuit current calculations for 3-phase systems referred to as the point-to-point method. Two examples of using this calculation method are provided.

SHORT-CIRCUIT CALCULATION REQUIREMENTS

Knowledge of and the ability to calculate the short-circuit current at various points in an electrical system is required to comply with various sections of OSHA regulations, the *NEC*, and *NFPA 70E*. The following list details a number of those requirements and explains why the short-circuit current is needed for compliance:

- A condition of use for determining the arc flash boundary per 130.5(B)(2) and using the arc flash PPE categories method (130.5(C)(2)) is that the available short-circuit current must not exceed Table 130.7(C)(15)(A)(b) maximum short-circuit current available parameter values.
- The procedure for calculating the arc flash boundary per 130.5(B)(1) and the incident energy per 130.5(C)(1) typically requires the available short-circuit current as part of the process.
- *NEC* 110.24: The maximum available fault current must be marked on service entrance equipment (with some exceptions) to help ensure the proper application of overcurrent protective device (OCPD) interrupting ratings and electrical equipment short-circuit current ratings.
- OSHA §1910.303(b)(4), *NEC* 110.9, and *NFPA 70E* 210.5: Overcurrent devices must have adequate interrupting ratings for the short-circuit current available.
- OSHA §1910.303(b)(5) and *NEC* 110.10: System components and assemblies must have adequate short-circuit current ratings for the short-circuit current available.
- *NEC* 240.12, 620.62, 645.27, 695.3(C), 700.28, 701.27, and 708.54: OCPDs must be selectively coordinated for the full range of overcurrents, including the available short-circuit current.

SHORT-CIRCUIT CURRENT CALCULATION BASICS

The need to determine available short-circuit currents may range from just one point in a system to all the critical points in the entire electrical system. The short-circuit current study for a system typically starts from the service point and extends to one point or all the points desired. Available short-circuit current may be calculated at the installation point of various equipment types, such as service equipment, switchboards, panelboards, transfer switches, motor control centers, disconnects, and motor starters. The magnitude of short-circuit current available at a specific location depends on two key factors: (1) the magnitude of short-circuit current the electrical energy

Figure 8-1. SHORT-CIRCUIT CURRENT

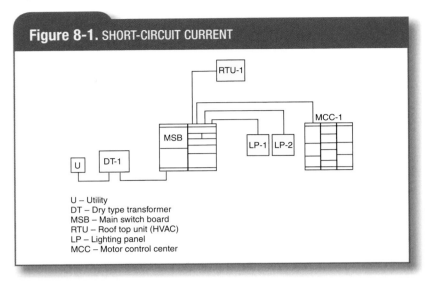

U – Utility
DT – Dry type transformer
MSB – Main switch board
RTU – Roof top unit (HVAC)
LP – Lighting panel
MCC – Motor control center

Figure 8-1. *Short-circuit current should be calculated at all critical points in the system.*

system, the greater the arcing fault current at that point.

Generally, for low voltage systems, the bolted short-circuit current determines the highest magnitude of current that the electrical system can deliver during a short-circuit condition. There are situations where line-ground short-circuit currents near the transformer secondary can be of a greater magnitude, but this will not be covered in this chapter.

For a system-wide study the short-circuit current calculations should be completed for all critical points in the system. **See Figure 8-1**. These would include, but are not limited to, the following:

- Switchboards
- Panelboards
- Motor control centers (MCC)
- Motor starters
- Disconnect switches
- Transfer switches

Sources of Short-Circuit Current

Sources of short-circuit current that are normally taken under consideration include utility generation, local generation, and synchronous and induction motors. In addition, alternative energy sources, such as batteries, photovoltaic systems, fuel cells, wind turbines, and flywheels, may also feed into an AC fault in some cases through a DC-to-AC converter. When electrical systems are supplied from a utility- or customer-owned transformer, the amount of short-circuit current depends on the size (kVA) and impedance (%Z) of the transformer. The

source(s) such as the utility system, generators, and running motors can deliver and (2) the impedance (resistance and reactance) of transformers, conductors, and busway from the source(s) to the point of the fault. In general, the short-circuit current decreases as distance from an electrical source increases due to the increased impedance added by such circuit components that have impedance, such as transformers, conductors, and busway.

Normally for arc flash risk assessments on low voltage systems, short-circuit studies involve calculating a bolted 3-phase short-circuit current condition. This scenario can be characterized as all three phases "bolted" together to create a zero-impedance connection. It establishes a faulted circuit condition that results in maximum thermal and mechanical stress in the system for evaluation of the OCPD interrupting rating and the equipment short-circuit current ratings. In addition, as covered in the chapter "Bolted and Arcing Fault Current and Reading Time-Current Curves," the arcing fault current magnitude depends on the magnitude of the available short-circuit current. The greater the available short-circuit current at a point in the

 Caution

Short-circuit currents can change over time due to system changes. To ensure protection of both people and equipment, short-circuit current studies and arc flash risk assessment studies must be updated as necessary.

larger the kVA rating and/or the lower the impedance of a transformer, the higher the available short-circuit current.

The available short-circuit current cannot be determined or estimated just by considering the ampere rating of an electrical panel or the type of facility. For instance, one 800-ampere distribution panelboard in a facility may have an available short-circuit current of 15,000 amperes at its line terminals, whereas another 800-ampere distribution panel in the same facility or in another facility may have an available short-circuit current of 80,000 amperes. Calculations should be done to determine the available short-circuit current.

Small residential building systems (100-A to 200-A service) typically have short-circuit currents of 10,000 amperes to 15,000 amperes or less. Small commercial building systems (400-A to 800-A service) typically have short-circuit currents of 15,000 amperes to 30,000 amperes. Larger commercial and manufacturing building systems (2,000-A to 3,000-A service) typically have short-circuit currents in the range of 50,000 amperes to 65,000 amperes. These short-circuit current values can be much higher where low-impedance transformers are used to increase efficiency. When commercial buildings are directly connected to a utility "low-voltage grid system," such as in major metropolitan cities (for example, New York, Chicago, Dallas), the short-circuit currents can exceed 200,000 amperes. The busway plug-in drops in auto production plants may have 150,000 amperes available short-circuit currents. Some hospitals with large alternate supply generators that can operate in parallel may have short-circuit current available exceeding 100,000 amperes.

Short-Circuit Current Factors

The magnitude of available short-circuit current at a specific point depends on many factors, including the short-circuit current contribution of the utility, local generation, and motors as well as the kVA rating and impedance of the trans-

Caution

Short-circuit currents can be very high (65,000 amperes to more than 200,000 amperes). For this reason, it is important to verify short-circuit current levels when applying overcurrent devices and equipment. It is also necessary to determine the available short-circuit current when doing an arc flash risk assessment.

former, the size and length of the wires, and more. **See Figure 8-2**.

The short-circuit current is typically the highest at the service point. Further in an electrical system, the circuit conductors and additional transformers will decrease

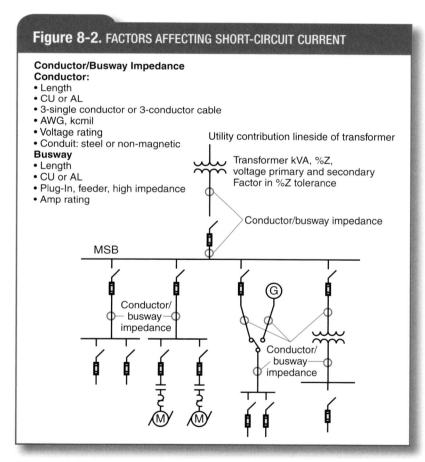

Figure 8-2. FACTORS AFFECTING SHORT-CIRCUIT CURRENT

Figure 8-2. Factors that affect the available short-circuit current include the utility, transformer (kVA and %Z), generators, motors, voltage, and conductor size and length. *Courtesy of Eaton's Bussmann Business.*

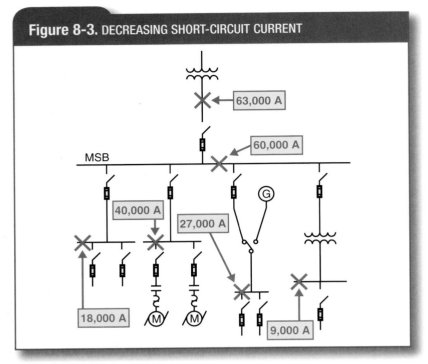

Figure 8-3. DECREASING SHORT-CIRCUIT CURRENT

Figure 8-3. The short-circuit current typically decreases downstream due to the presence of conductors and transformers. Courtesy of Eaton's Bussmann Business.

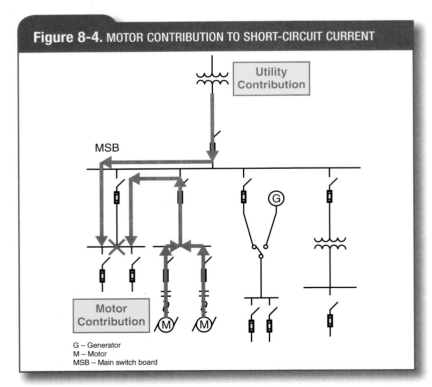

Figure 8-4. MOTOR CONTRIBUTION TO SHORT-CIRCUIT CURRENT

G – Generator
M – Motor
MSB – Main switch board

Figure 8-4. Motors that are in operation at the time a fault occurs elsewhere in the system can contribute to the fault current magnitude. Courtesy of Eaton's Bussmann Business.

the available short-circuit current at downstream equipment. **See Figure 8-3.**

Effects of Generators and Motors on Short-Circuit Current

Generators and motors in a premise system that are operating at the time of a fault condition can contribute short-circuit current in addition to the short-circuit current the utility delivers. Typically, the range of a short-circuit current from a generator is 7 to 14 times the full-load ampacity of the generator, with a general rule of thumb being 10 times. If the generator is operated in parallel with the utility during transfer (closed-transition transfer), the short-circuit current of the generator is added to the short-circuit current of the utility. If not, the generator short-circuit current is typically less than the normal source (utility) and is not used unless the Electrical Worker is examining arc-flash hazards when the system is under generator power.

When a fault occurs, a motor in operation acts like a generator and will contribute to the short-circuit current. **See Figure 8-4.** The motor contribution is usually four to six times the full-load ampacity of the motor. The additional motor contribution normally must be added to the utility contribution to determine the required ratings of overcurrent devices and equipment as well as to be considered in the various arc flash hazard analysis scenarios.

Effect of Transformers on Short-Circuit Current

Generally, for a typical low voltage electrical distribution system, the transformer is the single most important component affecting the available short-circuit currents throughout an electrical system. This relationship is especially notable for service equipment. It is important to realize that the available short-circuit current at the service may change, and often increase, due to changes in utility contribution or the transformer's kVA rating and percentage impedance (%Z). If the available short-circuit current changes at the service equipment, it will also change further downstream in the system.

Transformer percentage impedance is subject to manufacturing tolerances. The marked (nameplate) transformer impedance value may vary ±10% as per by UL Standard 1561. As a consequence, under fault conditions, the short-circuit current available on the secondary of a transformer can vary from −10% to +10% of what is calculated using the nameplate percent impedance. When evaluating OCPD interrupting ratings and equipment short-circuit current ratings, the transformer %Z multiplied by 0.9 should be used for the calculation of maximum available short-circuit current to represent the highest available short-circuit current possible due to a −10% transformer percentage impedance tolerance. However, when calculating short-circuits current for arc flash risk assessment purposes, it is recommended to calculate two different short-circuit currents on the secondary side of a transformer:

1. Use the −10% transformer percent impedance tolerance (%Z times 0.9) which will result in a higher short-circuit current.
2. Use the +10% transformer percent impedance tolerance (%Z times 1.1) which will result in a lower short-circuit current.

130.5 Information Note No. 2 provides the rationale for using both of these short-circuit values. If calculations of short-circuit currents are performed for locations further downstream in the system, two sets of values should be calculated for all other downstream points: one set using the transformer highest short-circuit current, and the other set using the transformer lowest short-circuit current.

The transformer size (kVA), impedance (%Z), and secondary voltage affect short-circuit current. **See Figure 8-5.** Assuming an "infinite" primary current, the short-circuit current at point A for a 5% impedance transformer with a secondary voltage of 480/277 volts is equal to the full-load ampacity of the transformer (601 amperes) × 100/%Z = (601 amperes) × (100/5), which equals 12,020 amperes (601 amperes × 20). This value may be 10% higher due to the tolerance of the transformer imped-

ance (± 10). For instance, the available fault current at point A would be calculated as 601 amperes × 100/(5 × 0.9) = 13,356 amperes. Continuing with the same example:

- If the transformer is replaced with a lower impedance transformer of 2% (to increase transformer efficiency), as shown at point B, the short-circuit current increases by 2.5 times (100/%Z increases from 20 to 50) to 30,050 amperes (601 A × 100/2 = 30,050 A).
- If the kVA is increased to 1,500 kVA (1,804 A full-load ampacity), as shown at points C and D, the short-circuit current triples to 36,080 A and 90,200 A, respectively.
- If the transformer secondary voltage is 208/120 V, as shown in points E and F, the available short-circuit currents are 83,280 A and 208,200 A, respectively.

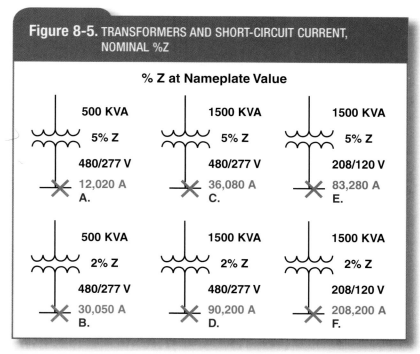

Figure 8-5. TRANSFORMERS AND SHORT-CIRCUIT CURRENT, NOMINAL %Z

% Z at Nameplate Value

500 KVA	1500 KVA	1500 KVA
5% Z	5% Z	5% Z
480/277 V	480/277 V	208/120 V
12,020 A A.	36,080 A C.	83,280 A E.
500 KVA	1500 KVA	1500 KVA
2% Z	2% Z	2% Z
480/277 V	480/277 V	208/120 V
30,050 A B.	90,200 A D.	208,200 A F.

Figure 8-5. Transformers have a significant effect on the available short-circuit current in an electrical distribution system and the variations in the short-circuit current for a transformer are based on transformer kVA, % impedance, and secondary voltage. Calculations on transformer secondary assume infinite primary available short-circuit current, nominal %Z, and no motor contribution.

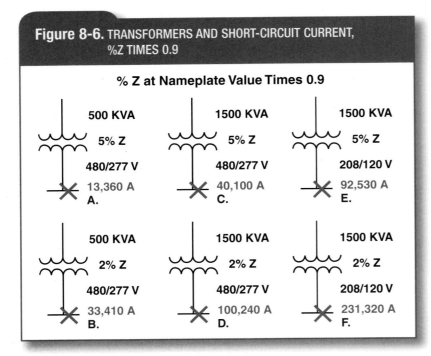

Figure 8-6. TRANSFORMERS AND SHORT-CIRCUIT CURRENT, %Z TIMES 0.9

% Z at Nameplate Value Times 0.9

500 KVA	1500 KVA	1500 KVA
5% Z	5% Z	5% Z
480/277 V	480/277 V	208/120 V
13,360 A **A.**	40,100 A **C.**	92,530 A **E.**

500 KVA	1500 KVA	1500 KVA
2% Z	2% Z	2% Z
480/277 V	480/277 V	208/120 V
33,410 A **B.**	100,240 A **D.**	231,320 A **F.**

Figure 8-6. Calculations on transformer secondary assuming infinite primary available short-circuit current, %Z times 0.9 representing −10% impedance tolerance. This produces the highest value for short-circuit current when considering only the transformer parameters.

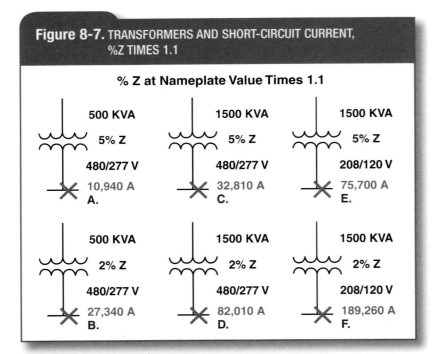

Figure 8-7. TRANSFORMERS AND SHORT-CIRCUIT CURRENT, %Z TIMES 1.1

% Z at Nameplate Value Times 1.1

500 KVA	1500 KVA	1500 KVA
5% Z	5% Z	5% Z
480/277 V	480/277 V	208/120 V
10,940 A **A.**	32,810 A **C.**	75,700 A **E.**

500 KVA	1500 KVA	1500 KVA
2% Z	2% Z	2% Z
480/277 V	480/277 V	208/120 V
27,340 A **B.**	82,010 A **D.**	189,260 A **F.**

Figure 8-7. Calculations on transformer secondary assuming infinite primary available short-circuit current, %Z times 1.1 representing +10% impedance tolerance, and no motor contribution. This produces the lowest value for short-circuit current.

 Tip

Transformer impedance percentage is subject to manufacturing tolerances. The marked impedance on a transformer nameplate may vary from plus or minus 10 percent from actual values determined by UL Standard 1561. Therefore, under fault conditions, the short-circuit current available on the secondary of a transformer can vary from −10 percent to +10 percent of what is calculated using the nameplate impedance percentage.

See Figure 8-6. These are the same transformers as used in Figure 8-5, but assuming the −10% tolerance transformer impedance. **See Figure 8-7.** These are the same transformers as used in Figures 8-5 and 8-6, but assuming the +10% tolerance transformer impedance. Note that calculated short-circuit current values can vary slightly depending on the calculation rounding method.

It is evident from the different available short-circuit currents shown that transformer characteristics have a major impact on the available short-circuit current. The transformer kVA rating and percentage impedance are very important for determining the available short-circuit current for the initial installation. However, it is also important to reassess the available short-circuit current if a transformer is replaced. If a utility transformer fails and the utility replaces the transformer, the percentage impedance can be higher or lower than the original by a significant amount. Most often, the percentage impedance of the newer transformers will be lower, resulting in a higher short-circuit current. Many utilities have a policy that if a transformer needs to be replaced due to failure, the utility can use a transformer with the next larger kVA rating. This practice is necessary in the event that the company's inventory of transformers with the same kVA rating is depleted at the time of replacement. Thus, the available short-

Background

Simple Method to Determine "Worst-Case" 3Ø Transformer Secondary Fault Current

This simple method can help to quickly assess a situation or to roughly double-check calculations. It assumes infinite available short-circuit current on the transformer primary. The easiest way to use this method:

Transformer Secondary Full-Load Amperes Known

Example transformer: 3Ø, 500 kVA, 480 V secondary, 2% Z (impedance)

$$I_{SCA} = \frac{\text{Trans FLA} \times 100}{\% Z}$$

$$= 601 \times \frac{100}{2}$$

$$= 601 \times 50$$

$$= 30,050 \text{ A Short-circuit current at transformer secondary}$$

As a memory aid, remember transformer full-load amperes for 500 kVA at 480 V and 208 V. The full-load amperes for 500 kVA transformer at 480 V is 601 A and at 208 V is 1,388 A. Then, for many situations other than 500 kVA, just use a multiple of these two. For instance, if the transformer is 1,000 kVA, just multiply by two: full-load amperes for 1,000 kVA transformer at 480 V is 2 × 601 = 1,202 A and at 208 V is 2 × 1,388 = 2,776 A.

If tolerance for %Z is desired to be used, then use (%Z × 0.9) for −10% tolerance or (%Z × 1.1) for +10% tolerance in place of the nameplate %Z.

Transformer Secondary Full-Load Amperes If Not Known

If the full-load amperes is not known, or remembered, then calculate the full-load amperes using the formula below and use the equation in the previous section. This is Step 1 from the 3-Phase Short-Circuit Calculation Procedure.

$$3\emptyset \text{ transf. } \quad I_{FLA} = \frac{\text{kVA} \times 1,000}{E_{L-L} \times 1.732}$$

circuit current could increase due to both a kVA rating increase and a percentage impedance decrease.

Effect of Conductors on Short-Circuit Current

The conductor size (or busway ampere rating), material (CU or AL), length, conduit magnetic or non-magnetic, and number per phase may also affect the short-circuit current. **See Figure 8-8.** Assuming a short-circuit current at point A of 40,000 amperes [at the beginning of the conductor run for each example] at 480/277 volts, the short-circuit current at the end of the conductor at point B would be 27,000 amperes. If the length of a 1 AWG conductor was increased to 50 feet, as shown from point A to C, the short-circuit current at

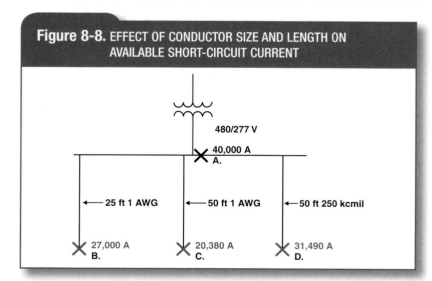

Figure 8-8. EFFECT OF CONDUCTOR SIZE AND LENGTH ON AVAILABLE SHORT-CIRCUIT CURRENT

Figure 8-8. The available short-circuit current at the end of a circuit conductor run is affected by the conductor's size and length.

point C would decrease to 20,380 amperes due to the increased impedance of the extra 25 feet of conductor. If the size of the conductor was then increased to 500 thousand circular mils (kcmil), as shown from point A to D, the short-circuit current at point D would be 31,490 amperes, which is higher than the resulting fault current for the 50 feet of 1 AWG at point C due to the decrease in the impedance per foot of the larger conductor.

EFFECT OF SHORT-CIRCUIT CURRENT ON ARC-FLASH RISK ASSESSMENTS

The available short-circuit current is a key factor that determines the extent of the arc-flash risk assessment. An integral part of protecting workers from an arc-flash hazard includes, but is not limited to, conducting an arc flash risk assessment. An arc flash risk assessment will determine, among other things, the arc flash boundary and incident energy or the required level of PPE using the arc flash PPE categories method.

When calculating the arc flash boundary (130.5(B)(1)) and using the incident energy analysis method (130.5(C)(1)), it is important to determine both the low and high values of short-circuit current that may occur at each point in the system. In some cases, a lower value of available short-circuit current means a lower arcing current may result in a longer clearing time of the overcurrent device. This longer clearing time may, in turn, cause a higher incident energy, greater AFB, and higher level of PPE required. See *NFPA 70E* 130.5 Informational Note No. 2. Lower arcing fault currents can result in increased opening times of OCPDs and higher incident energy compared to higher arcing fault currents, which can result in decreased OCPD opening time. **See Figure 8-9**. With both circuit breakers and fuses, a low arcing-fault current might potentially be in the OCPD's long time operating region, requiring many seconds to interrupt, which may result in the incident energy becoming greater than the incident energy resulting from a high available short-circuit current. Thus, the analysis should consider various electrical distribution system scenarios and variables to determine the lowest and highest available short-circuit current, with an incident energy analysis then being performed for each.

Scenarios Affect Arc Flash Risk Assessments

When conducting an arc flash risk assessment, the electrical distribution system may have variables that can result in different scenarios of available short-circuit currents, and therefore different scenarios necessary in determining the arc flash boundaries and incident energies or arc flash PPE categories. This subject will be covered more thoroughly in the chapter "Methods to Accomplish the Arc Flash Risk Assessment." Examples of the variable or scenarios include the following:

- Utility available
  - As accurate a possible
  - Infinite
- Transformer % impedance tolerance
  - Plus 10%
  - Minus 10%
  - Nominal marked value
- Switching scenarios: include multiple sources
  - Normal source
  - Onsite generation
  - Double ended substation with tie
  - Stored energy systems such as batteries
  - Redundant system designs
- Motor contribution
  - All running
  - None running

A switching scenario follows as an example to illustrate this concept.

Some facility's electrical distribution systems can be configured in various ways

Caution

Short-circuit currents can greatly increase when transformer impedance is decreased or when transformer kVA increases. When service transformers are changed, short-circuit studies and arc-flash studies must be updated as well. In addition, the interrupting ratings of OCPDs and short-circuit current ratings for equipment must be verified as sufficient: if these ratings are found to not be adequate, a safety hazard is present. Immediate action to remedy this problem should be taken.

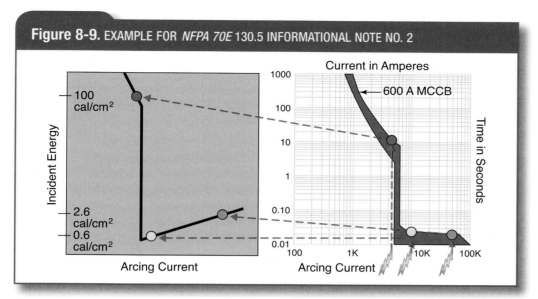

Figure 8-9. EXAMPLE FOR *NFPA 70E* 130.5 INFORMATIONAL NOTE NO. 2

Figure 8-9*. In some cases, a lower magnitude available short-circuit current results in lower magnitude arcing fault current, which causes significantly longer OCPD opening time and higher incident energy when compared to a higher available short-circuit current driving higher arcing-fault current, faster OCPD opening time, and associated incident energy.*

by opening or closing disconnects. For instance, a double-ended unit substation may include a low-voltage tie disconnect between the switchgear. One scenario might include both transformers energized and the low-voltage tie disconnect open between the switchgear. An alternative scenario might be the same as the first but with the tie disconnect closed. This second scenario might possibly have twice the magnitude of available short-circuit current at the switchgear as compared to the first scenario. **See Figure 8-10.** This variation can have a dramatic effect on determining the arc flash boundary (130.5(B)(1) or (2)), adhering to the condition of use parameters of the arc flash PPE categories method (130.5(C)(2) and Table 130.5(C)(15)(A)(b) parameters) or calculating the incident energy analysis method (130.5(C)(1)).

EFFECT OF SYSTEM CHANGES ON ARC-FLASH RISK ASSESSMENTS

Short-circuit currents can change over time due to changes in the electrical distribution system. If the electrical system is upgraded, expanded, or reduced, such

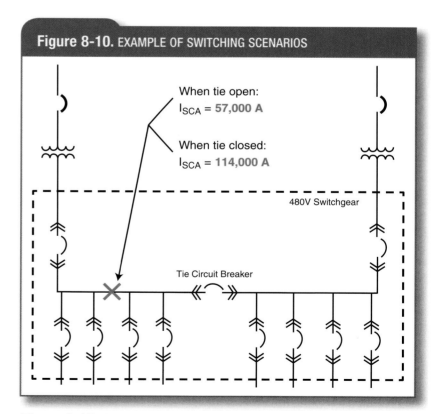

Figure 8-10. EXAMPLE OF SWITCHING SCENARIOS

Figure 8-10*. Double-ended unit substation with a tie circuit breaker can result in switching scenarios requiring calculating the available short-circuit current (I_SCA) and performing arc flash risk assessment for each.*

a change may affect the available short-circuit current throughout the system. Transformer changes were discussed earlier in this chapter. Other changes could include, but are not limited to, the utility system, generators, motors, or other building distribution system modifications. In recognition of these possibilities, *NFPA 70E* 130.5(2) requires the arc flash risk assessments to be reviewed whenever a major modification or renovation takes place. This information is also required to be reviewed periodically—no longer than every five years—to account for changes that may affect the results of the arc flash risk assessment.

PROCEDURES AND METHODS

To determine the short-circuit current at any point in the system, first draw a one-line diagram showing all of the sources of short-circuit current. Then include the system components, such as conductors and busway (size or ampere rating, lengths, material (CU or AL), number per phase, and conduit type or no conduit), and transformers (kVA, voltages, and percentage impedances). Information on the OCPDs (type, part number, ampere rating, voltage rating, interrupting ratings, and settings) is not necessary to perform a short-circuit study; however, this information is needed for an arc flash risk assessment. In order to perform a short-circuit current study, if data must be collected by examining the electrical equipment, it is suggested to record the necessary information to perform an arc flash risk assessment at the same time, including each OCPD (type, manufacturer, part number, ampere rating, and settings).

Short-circuit calculations are performed without OCPDs in the system. These calculations assume that these devices are replaced with copper bars, so as to determine the maximum "available" short-circuit current. After the calculation is completed throughout the system, current-limiting devices can be used to show reduction of the available short-circuit current at a single location only. Current-limiting devices do not operate in series to produce a compounding current-limiting effect.

Various methods have been developed to calculate the available short-circuit current, but all are based on Ohm's Law. These methods include the ohmic method, per-unit method, and point-to-point method, along with computer-based versions of all three methods. Several short-circuit calculation software programs are available. Some of these programs not only calculate the available

---

FC² Available Fault Current Calculator

There are online tools available to calculate short-circuit current. The FC² Available Fault Current Calculator can be accessed from the QR Codes shown. This application runs on Android and Apple devices or can be accessed online via the web.

For additional information, visit qr.njatcdb.org Item #1215.

For additional information, visit qr.njatcdb.org Item #1591.

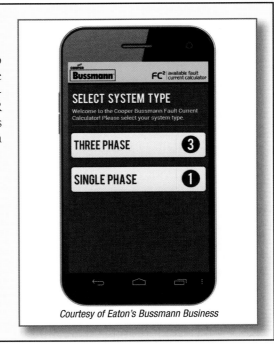

Courtesy of Eaton's Bussmann Business

short-circuit current, but also draw one-line diagrams, perform coordination studies, and determine the arc flash AFB and incident energy. These programs are available for purchase, and the price of the software program can vary greatly based on its capabilities.

The application of the point-to-point method permits the quick determination of available short-circuit currents with a reasonable degree of accuracy at various points for either 3-phase or single-phase electrical distribution systems. The procedures and examples in this chapter focus only on 3-phase systems. The point-to-point calculation methods for 3-phase and single-phase systems are available from Eaton's Bussmann Business in its Selecting Protective Devices (SPD) publication, which is available online at www.Bussmann.com. The point-to-point method is provided courtesy of Eaton's Bussmann Business.

POINT-TO-POINT METHOD

The point-to-point method uses equations to calculate the short-circuit current at:

- the secondary of a transformer assuming infinite primary short-circuit current available
- the end of a run of conductor or busway
- the secondary of a transformer with primary short-circuit current known.

Following the equations and informative notes shown here are two examples of how the point-to-point method is used.

At the end of this textbook are reference tables that are resources for using the point-to-point method. Review these tables along with their associated notes to become familiar with how this information is applicable in performing short-circuit current calculations. **See Appendix.**

3-Phase Short-Circuit Current Calculation Procedure

Equations to Calculate Short-Circuit Current on Transformer Secondary with Infinite Primary Available: Steps 1 to 3. Use the following procedure to determine the short-circuit current (I_{SCA}) at 3-phase transformer secondary terminals assuming infinite primary short-circuit current available:

Step 1: Determine the transformer full-load amperes (I_{FLA}) from either the nameplate, following equation, or **Figure A-1**.

$$3\emptyset \text{ transformer: } I_{FLA} = \frac{kVA \times 1000}{E_{L-L} \times 1.732}$$

Where:

E_{L-L} = line-to-line voltage
kVA = transformer kVA

Step 2: Determine the transformer multiplier

$$\text{Multiplier} = \frac{100}{\%Z}$$

Note 1: The marked (nameplate) transformer impedance (%Z) value may vary ±10% from the actual values determined by UL Standard 1561. For the high-fault condition scenario, multiply the transformer %Z by 0.9 and for the low-fault condition scenario, multiply the %Z by 1.1. Use 1.0 times %Z for no impedance tolerance affect.

Step 3: Determine the available short-circuit current (I_{SCA}) at 3-phase transformer secondary (**Figure A-6** provides a generalized estimate based on lowest %Z transformer survey.)

$3\emptyset$ transformer (see Note 2 for motor contribution): $I_{SCA\,(L-L-L)} = I_{FLA} \times \text{Multiplier}$

Note 2: The motor short-circuit contribution, if significant, may be determined and added to all fault locations throughout the system. A practical estimate of the motor short-circuit contribution is to multiply the total motor load current by 4 or 6. A good general rule is to calculate the total motor load current, multiply by 4, and add the result to step 3.

Note 3: The utility voltage may vary ±10% for power, and ±5.8% for 120 V lighting services. For worst-case conditions, multiply the values calculated in step 3 by 1.1 and/or 1.058, respectively.

Note 4: For 3-phase systems, line-to-line-to-line (L-L-L), line-to-line (L-L), line-to-ground (L-G), and line-to-neutral (L-N) bolted faults can be found in **Figure A-7** based on the calculated 3-phase line-to-line-to-line (L-L-L) bolted fault. Use IEEE 1584, *Guide for Performing Arc-Flash Hazard Calculations*, equations to calculate the 3-phase arcing current when performing an arc flash hazard analysis.

Equations to Calculate Short-Circuit Current at End of Conductor Run: Steps 4 to 6. Use the following procedure to determine the short-circuit current (I_{SCA}) at the end of a run of conductor or busway:

Step 4: Calculate the "f" factor

3ø line-to-line-to-line (L-L-L) fault

$$f = \frac{1.732 \times L \times I_{SCA\,(L\text{-}L\text{-}L)}}{C \times n \times E_{L\text{-}L}}$$

Where:

L = length (feet) of conduit to the fault
C = constant for conductors or busway (from **Figure A-2, A-3, or A-4**)
n = number of conductors per phase (adjusts the "C" value for parallel runs)
I_{SCA} = available short-circuit current in amperes (A) at the beginning of the circuit
$E_{L\text{-}L}$ = line-to-line voltage

Step 5: Determine the "M" (multiplier) from **Figure A-5** or the following equation:

Equation: $M = \dfrac{1}{1 + f}$

Step 6: Calculate the available short-circuit current (I_{SCA}) at the end of the circuit

3ø line-to-line-to-line (L-L-L) fault (see Note 2 for motor contribution):

$$I_{SCA\,(L\text{-}L\text{-}L)} = I_{SCA\,(L\text{-}L\text{-}L)} \times M$$
AT END OF CIRCUIT   AT BEGINNING OF CIRCUIT

Note 5: See Note 4.

Determining Short-Circuit Current on 3-Phase Transformer Secondary with Primary Short-Circuit Current Known: Steps A to C. When the primary short-circuit current at a transformer is known, either through a previous calculation (if it is the second transformer in the system) or if the short-circuit current of the utility connection is provided, then a more accurate calculation can be per-

---

Step A: Calculate the "f" factor for the transformer

3ø transformer: $f = \dfrac{I_{P(SCA)\,L\text{-}L\text{-}L} \times V_P \times 1.732 \times \%Z^*}{100,000 \times kVA}$

*Transformer Z is multiplied by 0.9 to establish the high-fault condition scenario. See Note 1.

Where:
$I_{P(SCA)}$ = primary short-circuit current
V_P = primary voltage (L-L)
kVA = transformer kVA

Step B: Determine the multiplier "M" for the transformer from **Figure A-5** or the following equation:

$M = \dfrac{1}{1 + f}$

Step C: Determine the short-circuit current at the transformer secondary (see Note 2 for motor contribution):

$I_{S(SCA)} = \dfrac{V_P}{V_S} \times M \times I_{P(SCA)}$

Where:
$I_{S(SCA)}$ = secondary short-circuit current
V_S = secondary voltage (L-L)

To determine the available short-circuit current at the end of a conductor or busway run connected to the secondary of this transformer, follow Steps 4, 5, and 6.

The point-to-point method courtesy Eaton's Bussmann Business.

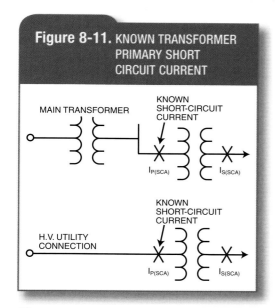

Figure 8-11. KNOWN TRANSFORMER PRIMARY SHORT CIRCUIT CURRENT

Figure 8-11. Use Steps A, B, and C when the available short-circuit current on the transformer primary is known.

formed. Use the following procedure Steps A to C to calculate the short-circuit current at the secondary when the short-circuit current at the transformer primary is known. **See Figure 8-11.**

Point-to-Point Method Examples

The following examples show how to use the point-to-point method equations and notes for 3-phase systems.

Reference tables are needed for some value, such as the conductor C values. **See Appendix.** For simplicity, the motor contribution and voltage variance are not included. See Notes 2 and 3 in the section 3-Phase Short-Circuit Current Calculation Procedure.

Example 1. This example has three points in the system to calculate the available short-circuit current. **See Figure 8-12.** This example assumes the utility available on the transformer primary is infinite.

CALCULATION EXAMPLE FOR FAULT #1

Step 1: Determine the transformer full-load amperes (I_{FLA})

$$I_{FLA} = \frac{kVA \times 1000}{E_{L\text{-}L} \times 1.732} = \frac{300 \times 1000}{208 \times 1.732} = 833 \text{ A}$$

Step 2: Find the transformer multiplier

$$\text{Multiplier} = \frac{100}{{}^{*}0.9 \times \text{Transf. \%Z}} = \frac{100}{1.8} = 55.55$$

*Transformer Z is multiplied by 0.9 to establish a high fault condition scenario. See Note 1 in the section 3-Phase Short-Circuit Current Calculation Procedure.

Step 3: Determine the short-circuit current (I_{SCA}) at the transformer secondary

$${}^{**}I_{SCA\,(L\text{-}L\text{-}L)} = I_{FLA} \times M = 833 \times 55.55 = 46,273 \text{ A}$$

**For simplicity, the motor contribution and voltage variance were not included. See Notes 2 and 3 in the section 3-Phase Short-Circuit Current Calculation Procedure.

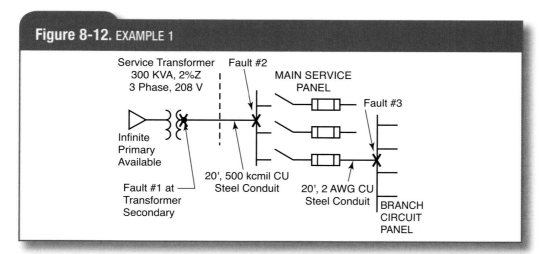

Figure 8-12. EXAMPLE 1

Figure 8-12. This one-line diagram provides information for Example 1, which has three short-circuit currents to be calculated.

CALCULATION EXAMPLE FOR FAULT #2
(Use $I_{SCA\ (L\text{-}L\text{-}L)}$ at Fault #1 in Steps 4 and 6.)

Step 4: Calculate the "f" factor

$$f = \frac{1.732 \times L \times I_{SCA\ (L\text{-}L\text{-}L)}}{C \times E_{L\text{-}L}} = \frac{1.732 \times 20 \times 46,273}{22,185 \times 208} = 0.35$$

Step 5: Determine the "M" (multiplier)

$$M = \frac{1}{1 + f} = \frac{1}{1 + 0.35} = 0.74$$

Step 6: Calculate the available short-circuit current (I_{SCA}) at Fault #2

$$I_{SCA\ (L\text{-}L\text{-}L)} \underset{\text{AT END OF CIRCUIT}}{} = I_{SCA\ (L\text{-}L\text{-}L)} \underset{\text{AT BEGINNING OF CIRCUIT}}{} \times M$$

$$= 46,273 \times 0.74 = 34,242\ A$$

CALCULATION EXAMPLE FOR FAULT #3
(Use $I_{SCA\ (L\text{-}L\text{-}L)}$ at Fault #2 in steps 4 and 6.)

Step 4: Calculate the "f" factor

$$f = \frac{1.732 \times 20 \times 34,242}{5,907 \times 208} = 0.97$$

Step 5: Determine the "M" (multiplier)

$$M = \frac{1}{1 + 0.97} = 0.51$$

Step 6: Calculate the available short-circuit current (I_{SCA}) at Fault #3

$$I_{SCA\ (L\text{-}L\text{-}L)\ AT\ FAULT\ \#3} = 34,242 \times 0.51 = 17,463\ A$$

Example 2. The following is an example of the procedure for the calculation of short-circuit current on a 3-phase transformer secondary with the primary short-circuit current known. **See Figure 8-13.**

Step A: Calculate the "f" factor for the transformer

$$f = \frac{30,059 \times 480 \times 1.732 \times 0.9^* \times 1.2}{100,000 \times 225} = 1.2$$

*Transformer Z is multiplied by 0.9 to establish a high-fault condition scenario. See Note 1 in the section 3-Phase Short-Circuit Current Calculation Procedure.

Step B: Determine the multiplier "M" for the transformer

$$M = \frac{1}{1 + 1.2} = 0.45$$

Step C: Determine the short-circuit current at the transformer secondary

$$I_{S(SCA)} = \frac{480}{208} \times 0.45 \times 30,059 = 31,215\ A$$

Figure 8-13. EXAMPLE 2

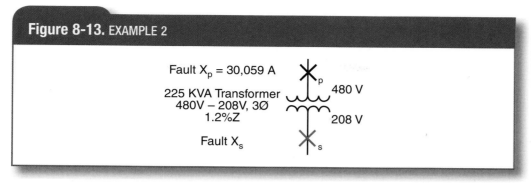

Fault X_p = 30,059 A

225 KVA Transformer
480V – 208V, 3Ø
1.2%Z

480 V

208 V

Fault X_s

Figure 8-13. *This one-line diagram provides information for Example 2, which has a 3-phase short-circuit current to be calculated on the secondary of a transformer when the available short-circuit current on the primary is known.*

Summary

Conducting a short-circuit current study is crucial in increasing electrical safety and is necessary to comply with the requirements of the *NEC* and *NFPA 70E*. This analysis helps ensure that the interrupting rating of overcurrent devices and the short-circuit current rating of equipment are adequate. In addition, a short-circuit study must be completed in order to perform an arc flash risk assessment. It is an important step to improve workplace safety by determining the AFB and the necessary PPE.

Short-circuit current calculations help the Electrical Worker to determine the short-circuit current, which can vary depending on the utility, transformers, generators, motors, voltage, conductors, and busway. Becoming familiar with the equations, tables, and notes associated with the point-to-point short-circuit calculation method described in this chapter will enable the Electrical Worker to make the proper calculations and use them to ensure proper protection of equipment and people.

Review Questions

1. The purpose of a short-circuit current study is to determine the available short-circuit current at one or many points in an electrical __?__ so as to ensure safety and compliance with OSHA, *NEC*, and *NFPA 70E* requirements.
   a. battery
   b. field
   c. generator
   d. system

2. The maximum available fault current __?__ be marked on new service entrance equipment (with some exceptions) to help ensure the proper application of OCPD interrupting ratings and electrical equipment short-circuit current ratings.
   a. cannot
   b. might
   c. must
   d. should

3. When doing an arc flash risk assessment using the incident energy method, the analysis should consider various electrical distribution system scenarios and variables to determine the __?__ available short-circuit current, with an incident energy analysis then being performed for that/those value(s).
   a. average
   b. highest
   c. lowest
   d. lowest and highest

4. Generally, the bolted short-circuit current determines the __?__ amount of current that the electrical system can deliver during a short-circuit condition (there are a few exceptions).
   a. fastest
   b. highest
   c. minimum
   d. slowest

5. The actual transformer impedance (%Z) value may vary ±10% from the marked (nameplate) value and therefore, for the high-fault condition scenario, multiply the transformer %Z by __?__.
   a. 0.1
   b. 0.9
   c. 1.0
   d. 1.1

6. The short-circuit current is typically the highest at the __?__.
   a. branch circuit panel
   b. feeder panel
   c. receptacle outlets
   d. service point

7. If a utility transformer fails and the utility replaces the transformer, most often the percentage __?__ of the newer transformers will be lower, resulting in a higher short-circuit current.
   a. impedance
   b. volt-amperes
   c. voltage
   d. wattage

8. Generally, the __?__ is the single most important component affecting the available short-circuit currents throughout an electrical system.
   a. distance from transformer to the service panel
   b. facility square footage
   c. largest motor
   d. transformer

9. Various methods have been developed to calculate the available short-circuit current, but all are based on __?__.
   a. *NFPA 70*
   b. *NFPA 70E*
   c. Ohm's Law
   d. OSHA regulations and standards

10. For arc flash risk assessment, whether using the incident energy analysis method (130.5(C)(1)) or the arc flash PPE categories method (130.5(C)(2)), typically the __?__ is a necessary value that must be determined.
    a. ambient temperature
    b. available short-circuit current
    c. distance to the exit door
    d. square footage of electrical room

Methods to Accomplish the Arc Flash Risk Assessment

Chapter Outline

- Preliminary Information
- Arc Flash Risk Assessment Methods
- IEEE 1584 AFRA Calculation Methods and Tools
- AFRA Equipment Labeling and Documentation
- Other Considerations

Chapter Objectives

1. Understand the concepts of protecting-OCPD and condition of maintenance when determining an arc flash boundary or determining the arc flash PPE using either the incident energy analysis method or the arc flash PPE categories method.
2. When using the arc flash PPE categories method, understand the parameters of Table 130.7(C)(15)(A)(b).
3. When an incident energy analysis method is used, understand the concepts and analysis steps.
4. Understand there are other considerations for an arc flash risk assessment and other related labeling requirements.

Chapter 9

Reference

1. *NFPA 70E®*, 2015 Edition

An electrician died as a result of burns he received while he was making a connection on a circuit breaker. The victim worked for an electrical contractor that employed 130 people. Safety functions were assigned to the firm's two project managers as an additional duty. The firm had no written safety policy or established safety program, and only on-the-job training was provided. The managers met with the foremen monthly to discuss safety issues.

On the day of the accident, the victim and his helper were to install three new circuit breakers into a panelboard that supplied power to various parts of an industrial complex. Three additional circuit breakers had previously been installed; they supplied power to the occupied portion of the building. The circuit breakers to be installed on the day of the accident were expected to control systems in the unoccupied portion of the building.

Before beginning the job, the victim was instructed by the foreman to install the top breaker first and the bottom breaker last. Although the victim had worked for the firm for only four days, he had approximately 16 years of experience as an electrical worker.

The victim did not follow the foreman's instructions. He secured the three breakers in the panelboard, and began to wire the bottom breaker first. The wires to be connected to the breakers were fed through a conduit in the bottom of the panelboard. The wires then had to be fed to the upper portion to make the connections. The line side of the bottom breaker became energized once the connections were completed. The victim then began to make the connections on the middle breaker without replacing the cover on the bottom breaker. He started to feed the wires for the middle breaker through the panelboard. He successfully fed four of the six wires to the upper portion. As the fifth wire was being fed upward, its uninsulated tip contacted one of the exposed energized connection points. The contact caused an arc, and the bottom breaker exploded. The victim was standing immediately in front of the breaker. His clothes caught fire, and he received massive burns to the upper portion of his body.

When the helper heard the explosion, he ran. As he was running away, a second explosion occurred. The helper received second- and third-degree burns to his left shoulder and back and to the back of his left arm.

The foreman was headed toward the room when he heard the explosions. He found the victim lying on the floor with his clothes on fire. He pulled the victim outside the room and extinguished the flames with his hands, receiving second- and third-degree burns in the process. The victim was conscious and talking at this time. The victim received burns to 80% of his body. Emergency rescue transported the victim to the hospital. He was later transferred to a burn center, where he died the following morning. The cause of death was attributed to massive burns to the victim's upper body.

Source: For details of this case, see FACE 8644. Accessed September 14, 2012.

For additional information, visit qr.njatcdb.org Item #1198.

INTRODUCTION

An earlier chapter, *Justification, Assessment, and Implementation of Energized Work,* covers the requirements of 130.5, Arc Flash Risk Assessment. This chapter focuses on an expanded discussion of the methods to determine the arc flash boundary and incident energy exposure level or arc flash PPE category. Examples are provided, including labeling and a discussion of related topics. Material in earlier chapters should be understood to comprehend and apply the material in this chapter, including available short-circuit currents, arcing fault currents, and determining overcurrent protective device (OCPD) clearing times.

PRELIMINARY INFORMATION

Prior to covering these topics the subjects of "condition of maintenance" and "protecting-OCPD" will briefly be discussed since both of these topics are relevant to determining the arc flash risk assessment.

Protecting-OCPD

The arc flash risk assessment determines the arc flash boundary (AFB) and appropriate arc flash PPE when there is an arc flash hazard. The arc flash hazard that may occur during an arc flash event is directly dependent on the OCPD that is protecting the equipment. This chapter will use the term "protecting-OCPD" for this OCPD. **See Figure 9-1.** "Protecting-OCPD" is a phrase used in this chapter; however, this phrase is not used in *NFPA 70E*. Generally, an arc flash risk assessment is determined at a specific piece of electrical equipment, such as a switchboard, panelboard, motor control center, or industrial control panel. The protecting-OCPD is always on the line side or upstream of the potential arcing fault location, and in most cases, this OCPD is in a separate enclosure.

Condition of Maintenance

NFPA 70E Chapter 1 has requirements for considering maintenance in 110.1(B) and other sections which will be mentioned

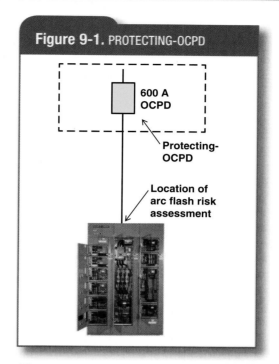

Figure 9-1. PROTECTING-OCPD

600 A OCPD

Protecting-OCPD

Location of arc flash risk assessment

Figure 9-1. This illustrates the concept of protecting-OCPD.

in the next two sections of this chapter. In addition, Chapter 2 of *NFPA 70E* is titled *Safety Related Maintenance Requirements.* See the chapter *Maintenance Considerations and OCPD Work Practices* for more information on maintenance and *70E* maintenance requirements.

When conducting an arc flash risk assessment in accordance with 130.5, the "condition of maintenance" may affect the assessment results in at least two ways.

Protecting-OCPD Condition of Maintenance. Per 130.5(3) the arc flash risk assessment must take into consideration the protecting-OCPD condition of maintenance. 130.5 Informational Note No. 1 provides the rationale; improper or inadequate maintenance may result in the protecting-OCPD taking longer to clear the fault current than its specified published clearing time. As a consequence, an actual arc flash incident may release significantly greater thermal energy than

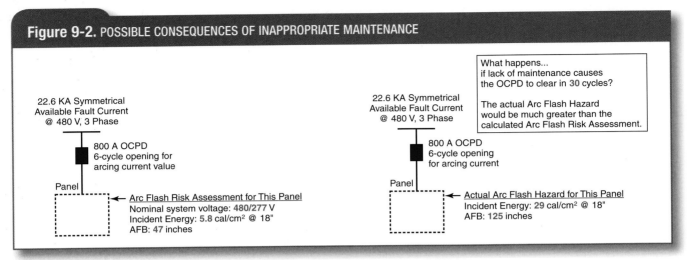

Figure 9-2. POSSIBLE CONSEQUENCES OF INAPPROPRIATE MAINTENANCE

Figure 9-2. The arc-flash risk assessment versus actual hazard that could occur due to poor or no maintenance is illustrated: Assuming the OCPD has been maintained and operates as specified by manufacturer's performance data, the incident energy would be 5.8 cal/cm² at 18 inches and the AFB is 47 inches. If lack of maintenance causes the OCPD to clear in 30 cycles rather than in six cycles, the actual incident energy is 29 cal/cm² at 18 inches, and the AFB would be 125 inches. The AFBs and incident energies were determined using IEEE 1584 equations.

that determined by the arc flash risk assessment.

See Figure 9-2. The 800A protecting-OCPD for the panel has a published clearing time of six cycles (0.1 second) for the calculated arcing fault current. The arc-flash risk assessment values shown were calculated using the published six-cycle clearing time for the protecting-OCPD. However, if the protecting-OCPD has not been properly maintained, it may take longer to interrupt the arcing fault current. As an example, if the OCPD takes 30 cycles to clear the arcing fault current due to poor maintenance, then the actual arc flash energy released will be greater than that determined by the arc flash risk assessment. This scenario illustrates one reason why maintenance of OCPDs is so important.

What if the condition of maintenance for a protecting-OCPD is questionable or known to be inadequate? While *NFPA 70E* 130.5(3) states to "take into consideration" the OCPD condition of maintenance, *NFPA 70E* does not provide specific requirements or actions to take if the protecting-OCPD is maintained in-

adequately or improperly. The circumstances of a specific task or work plan may affect the particular action to be taken. The risk assessment procedure requirements in 110.1(G) that must be part of an electrical safety program and the guidance provided in Informative Annex F, Risk Assessment Procedure, are among the *70E* provisions that provide a framework for how to take OCPD maintenance into consideration in the required arc flash risk assessment. See also 110.1(A) and 110.1(B) for additional insight into risk and condition of maintenance requirements. If the original objective was to perform justified energized work, it may be advisable to further investigate the work plan and establish an electrically safe work condition rather than work on the equipment energized. However, the Electrical Worker still has to interface with the equipment to check for the absence of voltage in the process of putting the equipment in an electrically safe work condition. If the task is to open a large ampere rated disconnect or rack-out a circuit breaker, using a remote operator or remote racking device, or utilizing an upstream disconnecting means that has been properly maintained, may be an approach to consider.

When the protecting-OCPD condition of maintenance is unacceptable, other scenarios may need to be explored. For example, a maintenance manager for a large automotive industrial facility has a policy to investigate upstream OCPDs to locate an OCPD device which has a suitable condition of maintenance and use that device for the arc flash risk assessment. With this approach, the AFB and arc rated PPE determined will have greater values than if the nearest upstream OCPD that has not been properly maintained is used for the assessment. Professional services companies that routinely perform AFB calculations, incident energy analysis and equipment labeling for facilities, generally have disclaimers that the calculation results are based on OCPDs performing to their published time-current characteristics. Some of these professional service companies will not use an OCPD in their calculations if it is obvious the OCPD has not been properly maintained or if an old technology is problematic, such as a circuit breaker that utilizes oil dashpot overcurrent sensing technology.

Per the last paragraph of 130.1, the condition of maintenance of the protecting-OCPD applies in circumstances such as determining the AFB (130.5(A)), calculating the incident energy (130.5(C)(1)), or determining the arc flash PPE category (130.5(C)(2)).

Equipment Condition of Maintenance.
In general, maintenance of all electrical equipment is important for electrical safety. It is also required to be considered as part of an electrical safety program per 110.1(B). When conducting an arc flash risk assessment, the condition of the electrical equipment with which the worker will interface is a factor for the risk assessment.

When using the arc flash PPE categories method, *NFPA 70E* Table 130.7(C)(15)(A)(a) has a column heading labeled "Equipment Condition*", which provides conditions of use as to whether arc flash PPE is required. The * in the heading directs the reader to a note at the end of the table which provides the meaning of "Equipment Condition" as "the equipment is properly maintained." So when using this Table, for a given task in the left column, whether arc flash PPE is required, in part, is dependent on whether the equipment is properly maintained or not. In this instance, the concern is the condition of maintenance of the equipment on which the worker will be working. Direction from the equipment manufacturer and consensus standards such as *NFPA 70B* and NETA are among the resources available for equipment maintenance guidelines. This condition of maintenance for example includes the equipment cleanliness, condition of conductor terminations, bus structure connection points, etc.

ARC FLASH RISK ASSESSMENT METHODS
Methods are available to determine the arc flash boundary (AFB) and arc flash PPE required by 130.5 Arc Flash Risk Assessment. There are methods for both ac and dc systems. It is important to recognize that each method has limitations and specific parameters for which they are applicable. These limitations/parameters may be such items as voltage, ac or dc, arc in a box, arc in open air, available short-circuit current, working distance. If a particular situation does not meet the limitation/parameters for a particular method, then that method cannot be used. Note that the arc flash PPE categories method is in the requirements of *NFPA 70E* while a number of incident energy calculation methods are shown in non-mandatory *NFPA 70E* Informative Annex D.

Arc Flash Boundary Determination
The AFB can be determined by calculating it per 130.5(B)(1), or, 130.5(B)(2) permits the AFB to be determined by using the arc flash PPE categories method.

When using the arc flash PPE categories method, the AFB is provided in Tables 130.7(C)(15)(A)(b) *Arc-Flash Hazard PPE Categories for Alternating Current (ac) Systems* or 130.7(C)(15)(B) *Arc-Flash Hazard PPE Categories for Direct Current (dc) Systems* via column "Arc Flash Boundary" as a fixed number. Determining the AFB by using the arc flash PPE categories method, as permitted in 130.5(B)(2), will be covered in this chapter when the arc flash PPE categories method 130.5(C)(2) is discussed. This AFB is valid where the conditions of use, including meeting the associated parameters, are met.

When using a calculation method, the AFB is the distance from the arcing fault source where the incident energy is 1.2 cal/cm². *NFPA 70E* Informative Annex *D Incident Energy and Arc Flash Boundary Calculation Methods* provides information on various calculation methods to determine the AFB. The most common calculation method used in the industry for 15 kV or less inside systems, is IEEE 1584 *Guide for Performing Arc Flash Calculations,* which is covered in *NFPA 70E* Informative Annex D, D.4. IEEE 1584 parameters of use include arc-in-a-box or arc-in-open air, 208 V to 15 kV, three-phase, and 50 HZ to 60 HZ systems. The calculation of the AFB will be covered in this chapter when the incident energy analysis method 130.5(C)(1) is discussed, since these calculations methods are closely related.

Arc Flash PPE Categories Method

This method can be used in lieu of an incident energy analysis for determining the AFB and the selection of arc flash PPE if specific conditions and parameters are met. Among the applicable *NFPA 70E* sections are 130.5(B)(2), 130.5(C)(2), 130.7(C)(15)(A) (ac systems), 130.7(C)(15)(B) (dc systems), Table 130.7(C)(15)(A)(a) (ac and dc systems), Table 130.7(C)(15)(A)(b) (ac systems), Table 130.7(C)(15)(B) (dc systems), and Table 130.7(C)(16).

Insight into the basis for the arc flash PPE categories method is provided by Informational Notes Nos. 1, 2, and 3 to 130.7(C)(15). Informational Note No. 1 clarifies that this method is generally an estimated exposure level determination and Informational Note No. 3 notes this is "based on the consensus judgment of the committee."

The requirements in 130.7(C)(15)(A) stipulate when the arc flash PPE categories method is permitted to be used for ac systems. The conditions of use are:

1. The task must be in Table 130.7(C)(15)(A)(a).
2. The estimated maximum available short-circuit current at the equipment to be worked on must be equal or less than shown in Table 130.7(C)(15)(A)(b) under the parameters for the type of equipment.
3. The fault clearing time has to be equal or less than the maximum fault clearing time in Table 130.7(C)(15)(A)(b) under the parameters for the type of equipment.
4. The working distance must not be less than that shown in Table 130.7(C)(15)(A)(b) under the parameters for the type of equipment.

Example 1. Assume the task is testing for the absence of voltage on a specific installed 480V panelboard during lockout/tagout. Referencing Table 130.7(C)(15)(A)(a), this task requires arc flash PPE. Next, for a 480V panelboard in Table 130.7(C)(15)(A)(b) the parameters are:

1. Maximum of 25kA short-circuit current available
2. Maximum of 0.03 sec (2cycles) fault clearing time
3. Working distance 18 inches

If the specific installed panelboard under consideration has an estimated available short-circuit current of 25kA or less and the fault current duration is 0.03 seconds or less and in addition, the working distance for this task is 18 inches or more, then the parameter conditions of use are satisfied. Then Table 130.7(C)(15)(A)(b) can be used and it states "2" under column Arc Flash PPE Category and "3 feet" under column Arc Flash Boundary, for this example. Next, go to Table 130.7(C)(16) for the selection of personal protective equipment for PPE category 2 (applicable for this example).

Figure 9-3. GENERALIZED CLEARING TIMES FOR 400-A CIRCUIT BREAKER FAMILIES AND FUSE FAMILIES

Type Device Grouped in "Families" for Generalized Clearing Times	Generalized Clearing Time (seconds)	Comments
Molded case circuit breakers	0.025[1]	Clearing times for some devices may be longer or shorter. Arc flash reduction maintenance system technology may reduce the time for worker safety.
Insulated case circuit breakers	0.05[1]	
Low voltage power circuit breaker with instantaneous trip	0.05[1]	
Low voltage power circuit breaker with short-time delay (STD) and no instantaneous trip	STD setting	Arc flash reduction maintenance system technology can reduce clearing time significantly for worker safety
Current-limiting fuse	0.004 to 0.008	Clearing times for UL Class CF, J, RK1, and RK5 fuses in their current-limiting range

1. Clearing times based on fault current within the instantaneous trip range.
Note: IEEE 1584, Edition 2002, Table 1 provides typical clearing times for circuit breakers. Clearing times for circuit breakers can be verified by consulting the manufacturer's literature.

Figure 9-3. Example of possible circuit breaker and current-limiting fuse families for generalized clearing times.

Recall that the arc flash PPE category is required to be on the equipment label per 130.5(D)(3)a and on the energized electrical work permit per 130.3(B)(2)(5)a. In addition, the results of the arc flash risk assessment must be documented per 130.5(A) and that the data which was used to support information put on the equipment label must be documented per 130.5(D) 2nd to last paragraph.

NFPA 70E does not define or provide guidance on the specific meaning of the phrase "estimated maximum available short-circuit current" or how to determine this value. To illustrate one possible way, the following Example 2 demonstrates how some in the industry use infinite available on the primary of a transformer to provide an estimated maximum available short-circuit current on the transformer secondary. In doing so, the result is a conservatively high value.

The protecting-OCPD needs to be evaluated as to whether the parameter maximum fault clearing time is satisfied. For this parameter also, *NFPA 70E* does not provide guidance of how to do so. Some in the industry use a concept of OCPD "families." This concept groups types of OCPDs that have similar fault clearing times into generalized families. For instance, the family of 400A or less thermal magnetic circuit breakers or the family of 400A or less current-limiting

fuses. **See Figure 9-3.** There likely are other methods which people in the industry use to evaluate the fault clearing time parameter.

NFPA 70E provisions in 130.7(C)(9)(b) address outer layers and requires that garments worn as outer layers over arc-rated clothing must also be made from arc-rated material. Courtesy of Salisbury by Honeywell.

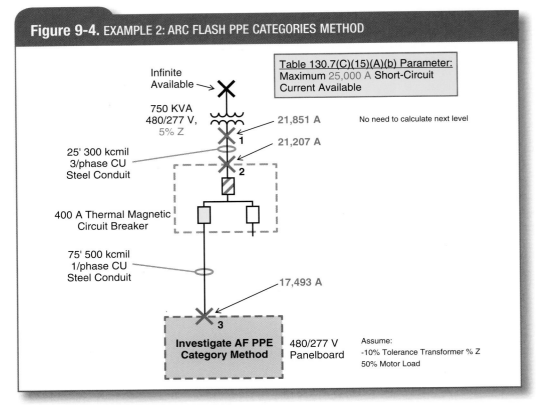

Figure 9-4. EXAMPLE 2: ARC FLASH PPE CATEGORIES METHOD

Infinite Available

Table 130.7(C)(15)(A)(b) Parameter:
Maximum 25,000 A Short-Circuit Current Available

750 KVA
480/277 V,
5% Z

21,851 A — No need to calculate next level

21,207 A

25' 300 kcmil
3/phase CU
Steel Conduit

1

2

400 A Thermal Magnetic
Circuit Breaker

75' 500 kcmil
1/phase CU
Steel Conduit

17,493 A

3

Investigate AF PPE Category Method

480/277 V
Panelboard

Assume:
-10% Tolerance Transformer % Z
50% Motor Load

Figure 9-4. *This is the single-line diagram and assumptions that are used to determine whether the estimated maximum fault available short-circuit current parameter is satisfied for an arc flash PPE categories method in Example 2.*

Example 2. See Figure 9-4. The 400-A panelboard has a branch circuit current monitoring system which needs troubleshooting while energized. To check the maximum available short-circuit current available: is X_3, the estimated maximum available short-circuit current at the panelboard, 25,000 A or less?

See Figure 9-5 for the output of the single-line diagram and results. Assumption of -10% tolerance for the transformer percent impedance (%Z) was made which provides a conservative short-circuit current value which can occur. Infinite available was assumed on the transformer primary. This will result in a conservative higher value. If the calculation at X_3 is close but too high, it may be worth the time to determine the available short-circuit current on the primary and redo the calculations using this

data. However, for a first pass, infinite available was assumed. The system has approximately 50% of the transformer rating utilized for motors. When a fault occurs on the 480-V system, running motors can contribute approximately 4 times their rated current to the faulted system and therefore need to be added.

To calculate X_3, first X_1 and X_2 must be calculated. The estimated available short-circuit current at X_1 is 21,851 A. Since this value is less than the parameter value for this situation, there is no need to calculate X_2 or X_3. The parameter is satisfied for this situation.

Example 3. This example is the same as Example 2 except the transformer has a 3.5 percent impedance rather than 5 percent. The objective in this example is to determine if the estimated maximum

Figure 9-5. EXAMPLE 2: ESTIMATED AVAILABLE SHORT-CIRCUIT CURRENTS

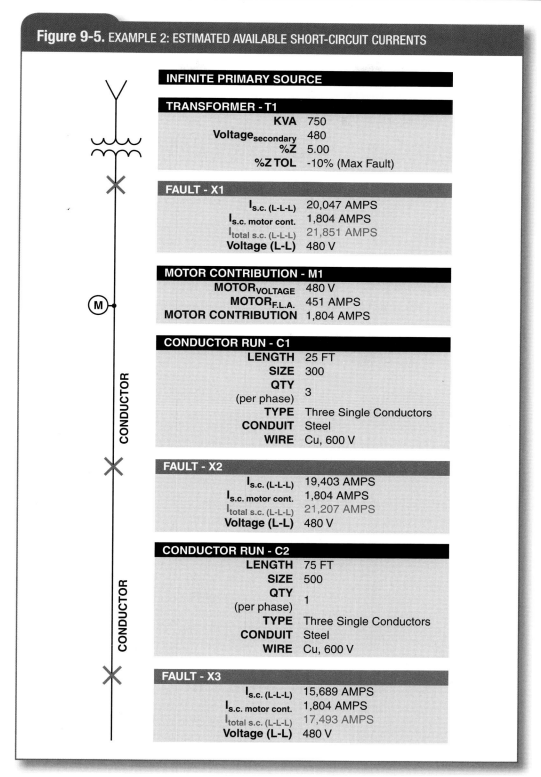

INFINITE PRIMARY SOURCE

TRANSFORMER - T1

KVA	750
Voltage$_{secondary}$	480
%Z	5.00
%Z TOL	-10% (Max Fault)

FAULT - X1

$I_{s.c. (L-L-L)}$	20,047 AMPS
$I_{s.c. motor cont.}$	1,804 AMPS
$I_{total s.c. (L-L-L)}$	21,851 AMPS
Voltage (L-L)	480 V

MOTOR CONTRIBUTION - M1

MOTOR$_{VOLTAGE}$	480 V
MOTOR$_{F.L.A.}$	451 AMPS
MOTOR CONTRIBUTION	1,804 AMPS

CONDUCTOR RUN - C1

LENGTH	25 FT
SIZE	300
QTY (per phase)	3
TYPE	Three Single Conductors
CONDUIT	Steel
WIRE	Cu, 600 V

FAULT - X2

$I_{s.c. (L-L-L)}$	19,403 AMPS
$I_{s.c. motor cont.}$	1,804 AMPS
$I_{total s.c. (L-L-L)}$	21,207 AMPS
Voltage (L-L)	480 V

CONDUCTOR RUN - C2

LENGTH	75 FT
SIZE	500
QTY (per phase)	1
TYPE	Three Single Conductors
CONDUIT	Steel
WIRE	Cu, 600 V

FAULT - X3

$I_{s.c. (L-L-L)}$	15,689 AMPS
$I_{s.c. motor cont.}$	1,804 AMPS
$I_{total s.c. (L-L-L)}$	17,493 AMPS
Voltage (L-L)	480 V

Figure 9-5. *This is the single-line diagram with estimated available short-circuit currents from the FC^2 Fault Current Calculator corresponding to Figure 9-4 for Example 2.* Courtesy of Eaton's Bussmann Business.

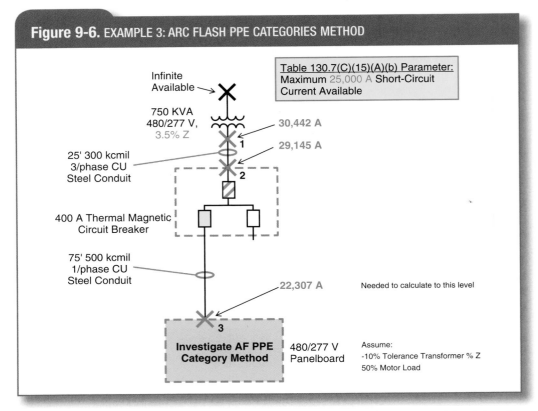

Figure 9-6. EXAMPLE 3: ARC FLASH PPE CATEGORIES METHOD

Table 130.7(C)(15)(A)(b) Parameter:
Maximum 25,000 A Short-Circuit
Current Available

Infinite Available

750 KVA
480/277 V,
3.5% Z

30,442 A

29,145 A

25' 300 kcmil
3/phase CU
Steel Conduit

400 A Thermal Magnetic
Circuit Breaker

75' 500 kcmil
1/phase CU
Steel Conduit

22,307 A

Needed to calculate to this level

**Investigate AF PPE
Category Method**

480/277 V
Panelboard

Assume:
-10% Tolerance Transformer % Z
50% Motor Load

Figure 9-6. This is the single-line diagram and assumptions that are used to determine whether the estimated maximum fault available short-circuit current parameter is satisfied for an arc flash PPE categories method in Example 3.

available short-circuit current at X_3 is equal or less than the parameter maximum of 25kA short-circuit current available. When doing the short-circuit current calculations, it is determined that the available short-circuit currents for X_1 and X_2 exceed the parameter maximum of 25,000 A. Therefore, the calculations should not stop at X_1 (or X_2). It is necessary to calculate X_3. In doing so, the calculation for X_3 results in 22,307 A estimated maximum available short-circuit current, which is less than the 25,000 A parameter value. **See Figure 9-6.**

AFB and Incident Energy Calculation Methods

NFPA 70E 130.5(B) and 130.5(C)(1) mandatory text does not provide the AFB or incident energy calculation methods (the equations). The most commonly used calculation methods are located in

NFPA 70E Informative Annex D, *Incident Energy and Arc Flash Boundary Calculation Methods*, which is provided for informational purposes only (nonmandatory). This Informative Annex contains useful information and references for these various calculation methods. The limitations/parameters for each of these calculation methods can be reviewed in *NFPA 70E* Informative Annex D, Table D.1 and in the Informative Annex text.

The 2002 IEEE 1584, *Guide for Performing Arc-Flash Hazard Calculations,* is the most up-to-date and widely used industry consensus standard on this topic for systems up to 15 kilovolts, especially for work involving enclosed electrical equipment that requires arc-in-the-box calculations, and is based on extensive arc fault testing and analysis. IEEE 1584 is the only calculation method which will be discussed in-depth.

IEEE 1584 AFRA CALCULATION METHODS AND TOOLS

The 2002 IEEE 1584 is based on extensive testing and analysis to develop equations to calculate the AFB, arcing current, and incident energy. Because it is the most up-to-date method, with the widest range of system characteristics and variables, it is considered by many to be the preferred calculation method for the AFRA. In addition, the arcing current equations in *NFPA 70E* Informative Annex D.4.2 and IEEE 1584 improve the ability to determine the arcing current. Note that these equations utilize units of J/cm^2 for incident energy and millimeters for distances. To convert units, use the following equivalencies: 1 cal/cm^2 = 4.184 J/cm^2 and 1 in. = 25.4 mm.

Refer to *NFPA 70E* Informative Annex D, D.4.1.1 for applicable system limitations for the IEEE 1584 basic equations methods. In summary 1584 is applicable to:

- 3-phase systems
- 208 V through 15 kV (theoretical equation above 15 kV)
- Short-circuit currents between 700 and 106,000 amperes
- Conductor gaps 13 mm through 152 mm
- Arcs-in-open-air or arcs-in-a-box

The 2002 IEEE 1584 includes three broad methods that are referred to in this chapter as basic equations, simplified fuse equations, and simplified circuit breaker equations, respectively. **See Figure 9-7.** In addition, equations for these methods have been incorporated into spreadsheets, programs, and commercial power system analysis packages resulting in simpler and quicker analysis results. Each of these methods will be discussed in the following sections.

For persons responsible for the calculations and for the workers relying on the calculated results to implement safe work practices and wear proper levels of arc rated PPE, it is important they understand this is not an exact science. There are many variables involved as well as dynamic, unpredictable behavior with arcing fault current. Even with the IEEE 1584 testing which was done under controlled and specific staged conditions, there are variations in the results. The exact same test run a few minutes later may have produced variations in the incident energy results. In the process of performing the arc flash risk assessment, the objective is to model the specific conditions and parameters of the actual circuit and equipment and then use the available IEEE 1584 equations with

Figure 9-7. IEEE 1584 ARC FAULT BOUNDARY AND ARC FLASH INCIDENT ENERGY METHOD COMPARISON

2002 IEEE 1584 Method	Brief Description (Key Differentiating Points)	Comments

Basic Equations	• Based on testing
	• Ability to change many variables
	• For any overcurrent protective device (OPCD)
	• Process:
	  1. Determine the available (bolted) short-circuit current
	  2. Use the IEEE 1584 arcing current equations to determine the arcing current
	  3. Using OCPD time–current characteristics, determine the OCPD clearing time for the arcing current
	  4. Repeat step 3 for 85% of arcing current
	  5. Use IEEE 1584 AFB and I.E. basic equations to calculate AFB and incident energy at specified working distance for both steps 3 and 4; select the worst case of the two results
		If doing manual calculations, this is the more tedious method
		Method suggested for specific circuit breaker if time–current curve is available
		Method suggested for a fuse that is not one where the simplified fuse equations are applicable

Figure 9-7. IEEE 1584 arc fault boundary and arc flash incident energy method comparison. This gives a high-level overview of the three methods.

the applicable available variables to get the closest match.

IEEE 1584 Basic Equations Method

The basic equation method includes the steps which are incorporated in Example 4. Consider the overcurrent protective device condition of maintenance per 130.5(3). If the OCPD providing the protection for arc flash hazard analysis has not been properly maintained, then the AFRA results of steps 5 and 6 may

not be reliable. Consider investigating alternate approaches.

Example 4. **See Figures 9-8 and 9-9.** This example will use the IEEE 1584 basic equation method to calculate the AFB (130.5(B)(1)) and incident energy (130.5(C)(1)) for a potential arcing fault in the panelboard shown in **Figure 9-8**. For each step there may be a brief discussion, referencing the specific equation(s) used in *NFPA 70E* Informative Annex D

Figure 9-8. EXAMPLE 4: SINGLE-LINE DIAGRAM

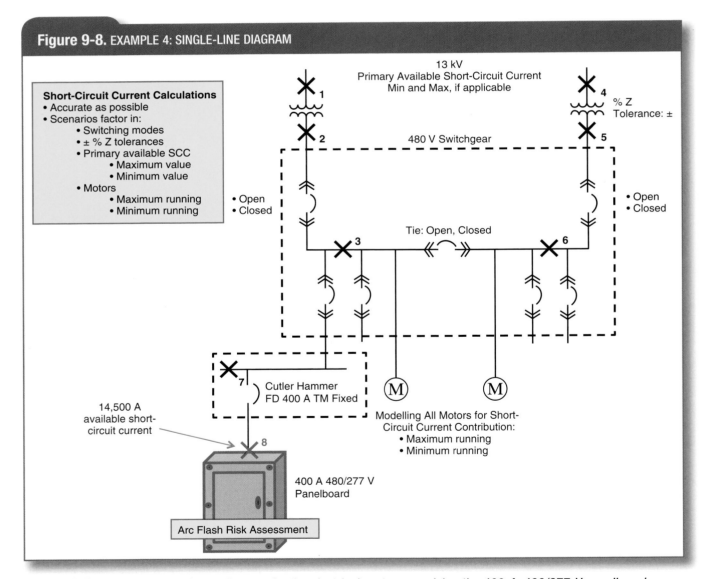

Figure 9-8. Example 4: single-line diagram for the electrical system supplying the 400-A, 480/277-V panelboard.

and mentioning the variable values for the equations that are used.

1. **Calculate the available (bolted) short-circuit current.**

   The single-line diagram shows potentially several different scenarios and calculations should be made for each scenario. For illustrative purposes, this discussion assumes only one scenario and **the available short-circuit current X_8, at the panelboard, is 14,500 A.** For any of the IEEE 1584 methods, the thoroughness and accuracy of the data and the short-circuit current calculations are important. This means having an accurate single-line diagram: up-to-date and accurate data of the system components, such as conductor sizes and lengths, full description of each OCPD, and transformer kVA and % impedance. This system has potentially several scenarios in which the available short-circuit current can vary. Steps 2 through 6 should be performed for each scenario and then the worst case AFB and incident energy should be chosen for the PPE selection and for the equipment labeling. Chapter *Fundamentals of 3-Phase Bolted Fault Current* provides information on various scenarios and how to perform a short-circuit current analysis. One professional services company, which routinely performs arc flash risk assessment studies for facilities, reported running 13 scenarios for a large, complex electrical system.

2. **Calculate the arcing current.**

   The calculation determines the arcing fault current as $I_a = 8,740$ A.

   This step is to determine I_a which is arcing current. In Informative Annex D, use equations D.4.2(a), which is applicable for system voltages less than 1 kV, and D.4.2(2). (D.4.2(b) is for systems 1 kV and greater systems.) The K variable is selected for arcs-in-a-box since this work is to be performed in a panelboard. G, the conductor gap variable, is selected from Table D.4.2 which shows 1 inch (25 mm) for panelboards. 1 inch

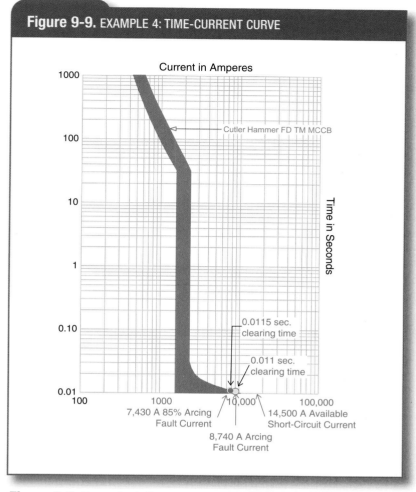

Figure 9-9. EXAMPLE 4: TIME-CURRENT CURVE

Figure 9-9. Example 4: time-current characteristic curve for Eaton Cutler Hammer FD fixed thermal magnetic molded case circuit breaker.

clearance through the air between phases or phase to ground is typical in UL Standards for 600-V rated panelboards, motor control centers, and similar equipment. 1½ inch clearance is typical for 600-V rated switchboards and switchgear.

3. **Determine the OCPD clearing time at the arcing current value.**

   Interpreting the time-current curve shown, this protecting-OCPD clears the 8,740 A arcing fault current in 0.011 second.

   For the IEEE basic equation method, the total or maximum clearing time-current characteristics of the specific protecting-OCPD should be used to provide accuracy: that is, manufacturer, type, ampere rating, and setting (if applicable). In Example 4, the circuit breaker is a Cutler Hammer 400-A FD fixed thermal magnetic molded case circuit breaker. Fixed means it's time-current characteristics are non-adjustable. **See Figure 9-9** which is the time-current curve for this specific circuit breaker.

4. **Repeat step 3 for 85% of the arcing current.**

   85% of the calculated arcing fault current is 7,430 A.

   The protecting-OCPD **clears this 7,430 A arcing fault current in 0.0115 second.**

   The IEEE 1584 testing demonstrated variance in the actual arcing current based on a test setup with a fixed calibrated available short-circuit current.

For situations where the arcing current happens to be in the steep (highly inverse) portion of an OCPD time-current curve, a small variance in arcing current can result in a large change in the incident energy and AFB. Therefore, as a precaution for the basic equation method, IEEE 1584 requires determining the AFB and incident energy at both the calculated arcing current and 85% of the calculated arcing current. Example 9 in this chapter illustrates why this is done.

5. **Calculate the incident energy**

   With 8,740 A arcing current and 0.011 second clearing time, the incident energy is calculated to be 0.32 cal/cm².

   With 7,430 A arcing current and 0.0115 second clearing time, the incident energy is calculated to be 0.28 cal/cm².

   In Informative Annex D, D.4.3: use equations D.4.3(a) and D.4.3(b) to determine E_n and then D4.3(c) to determine the incident energy, which is variable E. I_a represents the arcing current in the equations. The variable t is the clearing time of the protecting-OCPD at the arcing current. Calculations must be done using I_a as 100% arcing current with the associated OCPD clearing time t for that arcing current and then I_a equal to 85% arcing current with the associated clearing t for that arcing current.

6. **Calculate the AFB**

   Calculated AFB for 8,740 A arcing current and 0.011 second clearing time is 8 inches.

   Calculated AFB for 7,430 A arcing current and 0.0115 second clearing time is 7.4 inches.

   Informative Annex D, D.4.5: use equations D.4.3(a) and D.4.3(b) to determine E_n and then D4.5(a) to determine the AFB, which is variable D_B. I_a represents the arcing current in these equations. The variable t is the clearing time of the arc-flash-protecting-OCPD at the arcing current. Variable E_B repre-

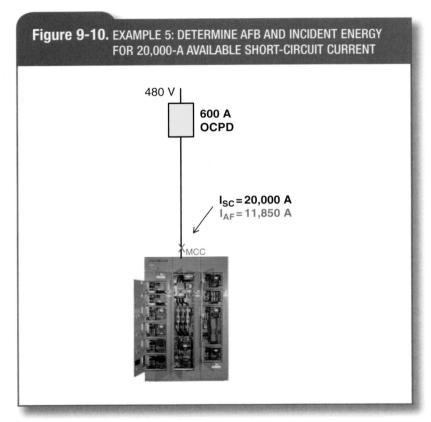

Figure 9-10. EXAMPLE 5: DETERMINE AFB AND INCIDENT ENERGY FOR 20,000-A AVAILABLE SHORT-CIRCUIT CURRENT

480 V

600 A
OCPD

I_{SC} = 20,000 A
I_{AF} = 11,850 A

MCC

Figure 9-10. An example for which the arc-flash incident energy is determined for protection by four different 600-A OCPD types; the motor control center has a 20,000-A available bolted short-circuit current, resulting in potentially an 11,850-A arcing fault current.

sents the thermal energy at the threshold of a second degree burn, which is 5 J/cm² (1.2 cal/cm²) per 130.5(B)(1). Calculations must be done for 100% arcing current and 85% arcing current and the associated clearing times of the OCPD at those arcing current levels, respectively.

Example 5: IEEE 1584 Basic Equation Method with Various OCPDs. This example demonstrates the results when using the AFB and incident energy calculation method for OCPDs with the same ampere rating but with different time-current characteristics.

Assume work must be done in the main disconnect section of a 600-A motor control center (MCC) when it is not in an electrically safe work condition. Also, assume that the bolted 3-phase short-circuit current at the line terminals of the MCC is calculated to be 20,000 amperes. Using the 2002 IEEE 1584 arcing fault equation for a 20,000 A available bolted short-circuit current, the arcing fault current is calculated as 11,850 A. 85% of the arcing current is 10,080 A. **See Figure 9-10.**

See Figures 9-11, 9-12, 9-13, and 9-14. The AFB and arc-flash incident energy is calculated for four different 600-A

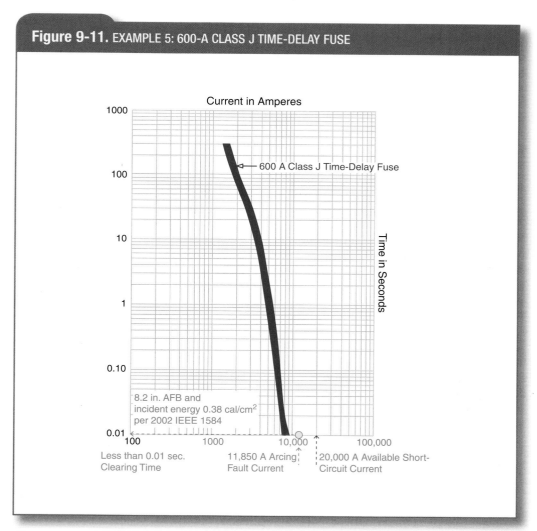

Figure 9-11. EXAMPLE 5: 600-A CLASS J TIME-DELAY FUSE

Figure 9-11. With a 20,000-A available short-circuit current, the arcing current is 11,850 A and the 600-A Class J time-delay fuse clears the arcing fault current in less than one-half cycle, resulting in an 8.2 in. AFB and incident energy of 0.33 cal/cm² at 18 in.

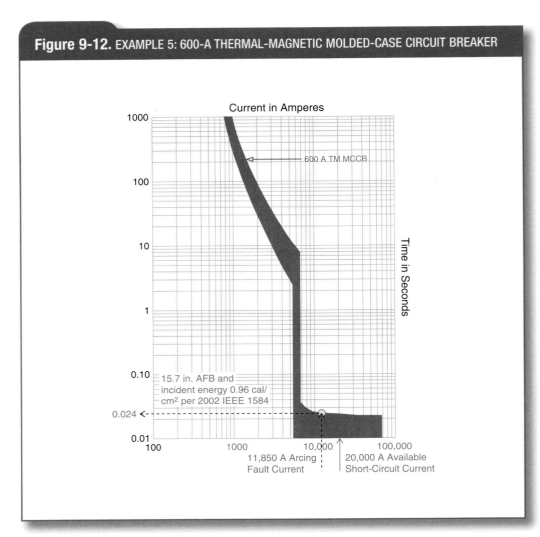

Figure 9-12. EXAMPLE 5: 600-A THERMAL-MAGNETIC MOLDED-CASE CIRCUIT BREAKER

Figure 9-12. With a 20,000-A available short-circuit current, the arcing current is 11,850 A and the 600-A thermal-magnetic molded-case circuit breaker clears the arcing fault current in 0.024 second, resulting in a 15.7 in. AFB and incident energy of 0.96 cal/cm² at 18 in.

Figure 9-13. EXAMPLE 5: 600-A ELECTRONIC MOLDED-CASE CIRCUIT BREAKER

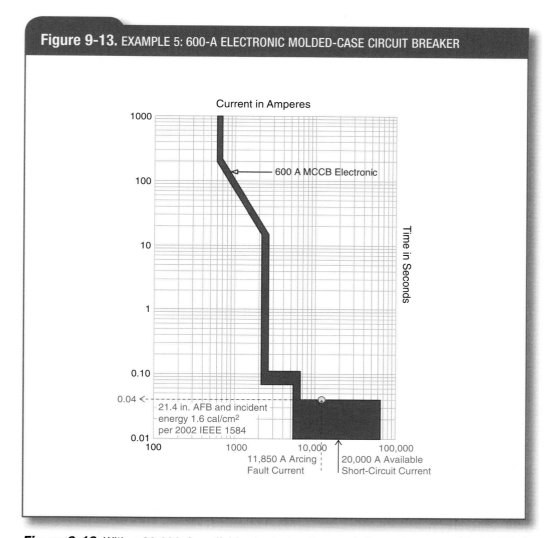

Figure 9-13. *With a 20,000-A available short-circuit current, the arcing current is 11,850 A and the 600-A electronic molded-case circuit breaker clears the arcing fault current in 0.04 second, resulting in a 21.4 in. AFB and incident energy of 1.6 cal/cm² at 18 in.*

Figure 9-14. EXAMPLE 5: 600-A LOW-VOLTAGE POWER CIRCUIT BREAKER

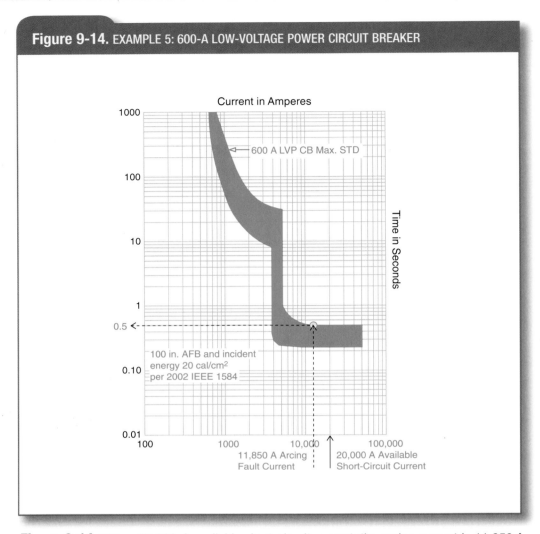

Figure 9-14. With a 20,000-A available short-circuit current, the arcing current is 11,850 A and the 600-A low-voltage power circuit breaker with short-time delay clears the arcing fault current in 0.5 second, resulting in a 100 in. AFB and incident energy of 20 cal/cm² at 18 in.

Figure 9-15. EXAMPLE 5: AFB AND INCIDENT ENERGY FOR 20,000-A AVAILABLE SHORT-CIRCUIT CURRENT

OCPD Type	OCPD Clearing Time for 11,850-A Arcing Fault Current	Arc Flash Boundary	Incident Energy at 18" Working Distance
600-A Class J fuse	½ cycle or less	8.2 in.	0.33 cal/cm²
600-A thermal-magnetic molded-case circuit breaker	0.024 second	15.7 in.	0.96 cal/cm²
600-A electronic-sensing molded-case circuit breaker	0.04 second	21.4 in.	1.60 cal/cm²
600-A low-voltage power circuit breaker with short-time delay set at 0.5 second	0.5 second	100 in.	20 cal/cm²

Figure 9-15. For the four different 600-A OCPD type scenarios, AFB and the incident energy at an 18-inch working distance can be determined using 2002 IEEE 1584 basic equations and the clearing times listed here.

OCPD types for the situation in **Figure 9-10** using 100% arcing fault current. The AFB and incident energies will not be shown for the calculations resulting from 85% of the arcing current. The reason is that for these cases, the resulting AFB and incident energy for 85% arcing current scenario was less than the results using 100% of the arcing current. IEEE 1584 basic equation method was used for the calculations.

First, the clearing time for each of the four different OCPD types is determined for the 11,850 A arcing current using the OCPD time–current curves. Then, by using the 2002 IEEE 1584 basic equations with these clearing times and the other parameters, the AFB and the incident energy at 18-inch working distance are determined for each. The AFB and incident energy for each of the four different OCPDs scenarios for 100% arcing current are shown in tabular format in **Figure 9-15**.

Example 6. **See Figure 9-16.** This is the same equipment layout as in Example 5 but the 3-phase available short-circuit current is 50,000 A resulting in 25,930 A arcing fault current. 85% arcing fault current is 22,040 A. **See Figure 9-17** for the tabular results of the AFB and incident energy analysis method of 130.5(C)(1). The clearing times for the four different OCPDs with 25,930 A arcing current can be determined from **Figures 9-11, 9-12, 9-13, and 9-14**. Compare the AFBs and

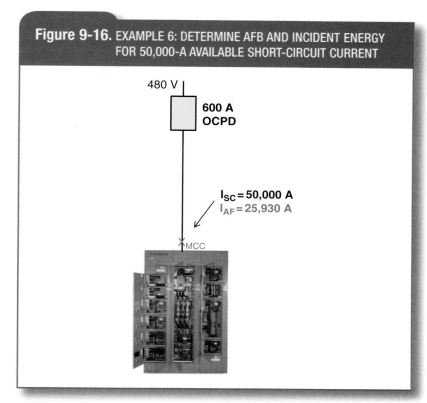

Figure 9-16. EXAMPLE 6: DETERMINE AFB AND INCIDENT ENERGY FOR 50,000-A AVAILABLE SHORT-CIRCUIT CURRENT

Figure 9-16. An example for which the arc-flash incident energy is determined for protection by four different 600-A OCPD types; the motor control center has a 50,000-A available bolted short-circuit current, resulting in potentially a 25,930-A arcing fault current.

incident energies of **Figure 9-15** with those in **Figure 9-17**. The resulting AFB and incident energy for the 85% arcing current scenario was less than the results using 100% of the arcing current in these cases, so these values were not shown.

OCPD Type	OCPD Clearing Time for 25,930-A Arcing Fault Current	Arc Flash Boundary	Incident Energy at 18" Working Distance
600-A Class J fuse	½ cycle or less	6 in.	0.25 cal/cm²
600-A thermal-magnetic molded-case circuit breaker	0.023 second	25.6 in.	2.14 cal/cm²
600-A electronic-sensing molded-case circuit breaker	0.04 second	35.9 in.	3.72 cal/cm²
600-A low-voltage power circuit breaker with short-time delay set at 0.5 second	0.5 second	167 in.	46.5 cal/cm²

Figure 9-17. EXAMPLE 6: AFB AND INCIDENT ENERGY FOR 50,000-A AVAILABLE SHORT-CIRCUIT CURRENT

Figure 9-17. For the four different 600-A OCPD type scenarios, AFB and the incident energy at an 18-inch working distance can be determined using 2002 IEEE 1584 basic equations and the clearing times listed here.

IEEE 1584 basic equation method was used for the calculations for the circuit breaker scenarios and IEEE 1584 simplified fuse equation method was used for the fuse scenario (simplified fuse equation method is discussed later in this chapter).

Example 7: Circuit Breakers with ARMS. A circuit breaker with short-time delay (STD) and no instantaneous trip holds off interrupting a fault until the STD time setting. **Figure 9-14** represents the time-current curve for an older style 600-A low voltage power circuit breaker with a STD set at 0.5 seconds (30 cycles). With an available short-circuit current of 50,000 A at 480 V, resulting in 25,930 A arcing fault current, the AFB is 167 in. and the incident energy is 46.5 cal/cm² at 18 in (reference the table in **Figure 9-17**). An arc flash reduction maintenance system (ARMS) could be retrofitted to this circuit breaker to reduce the arc flash hazard. **See Figure 9-18.** This ARMS can be switched on when a worker must work on the electrical equipment this circuit breaker is protecting and this specific ARMS has a clearing time of 0.04 seconds.

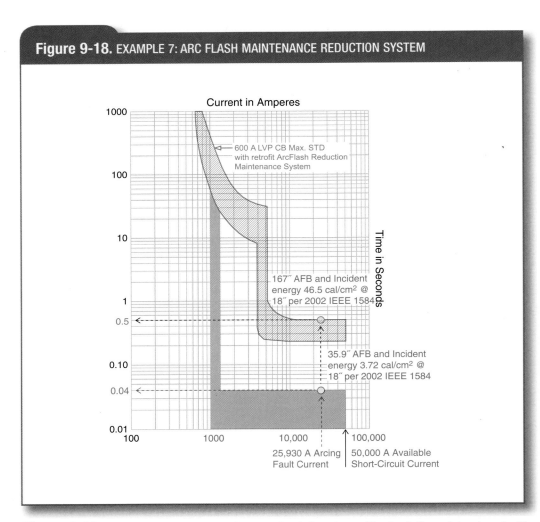

Figure 9-18. EXAMPLE 7: ARC FLASH MAINTENANCE REDUCTION SYSTEM

Figure 9-18. *Older style low voltage power circuit breaker with short-time delay is retrofit with an arcflash maintenance reduction system. This technology can greatly mitigate an arc flash hazard.*

Figure 9-19. EXAMPLE 7: COMPARING RESULTS OF FIGURE 9-14 TO FIGURE 9-18

Benefits of Arc Flash Reduction Maintenance Systems in Reducing Arc Flash Risk Assessment

	600-A LVPCB 30 Cycle Short-Time Delay	600-A LVPCB with ARMS turned "on" (0.04 clearing time)
AFB	167 in.	35.9 in.
Incident Energy @ 18 in.	46.5 cal/cm²	3.72 cal/cm²

Figure 9-19. Arc flash reduction maintenance systems can reduce the AFB and incident energy. This compares results from Figure 9-14 to Figure 9-18.

An assessment with the ARMS switched on results in an AFB of 35.9 in. and incident energy of 3.72 cal/cm². This is a dramatic reduction in the AFB and incident energy; see **Figure 9-19** for comparison. ARMS are available on new circuit breakers as well as can be retrofit to installed circuit breakers. The clearing time of ARMS vary significantly based on the manufacturer and technology used.

Example 8: Fusible Switches with ARMS. Large fusible circuits may result in large AFBs and high incident energies for lower magnitude arcing fault current levels (occurs with lower available short-circuit current systems relative to the OCPD ampere rating). Fusible switches equipped with arc flash maintenance re-

duction systems can reduce the arc flash risk assessment. This is combining two devices: the ARMS is best for arc flash incident energy reduction for lower arcing fault currents and the Class L fuses are best at reducing the arc flash incident energy for high arcing fault current.

See Figure 9-20. This table is the result of performing the IEEE 1584 AFB and incident energy calculations for Class L fuses by themselves and then the ARMS alone. The column on the far right is the resulting AFBs and incident energies for the combination of the fuses and switch working together; the ARMS does best at the lower fault current levels and the fuses do best at the higher fault levels. In combination, the AFB and incident energies are held to relatively low levels.

Figure 9-20. EXAMPLE 8: LARGE FUSIBLE SWITCH AND ARMS ANALYSIS

Available Short-Circuit Current	1200 A Class L Fuses		ARMS "on"		Resulting Best of Combination	
	AFB (inches)	Incident Energy (cal/cm² @ 18 in.)	AFB (inches)	Incident Energy (cal/cm² @ 18 in.)	AFB (inches)	Incident Energy (cal/cm² @ 18 in.)
10,000 A	>100	>120	16.6	1	16.6	1
20,000 A	92	17.4	24.5	2	24.5	2
50,000 A	8	0.39	41	4.7	8	0.39
100,000 A	8	0.39	61	8.8	8	0.39

Figure 9-20. For large ampere rated switches (1,200 A and greater) the addition of arc flash reduction maintenance system can be beneficial.

Range Where ARMS is Faster: See Figure 9-21. With 20,000 A short-circuit current, the arcing fault current is calculated as 11,850 A and the 85% arcing fault current is 10,080 A. With the clearing times shown, the 1,200-A Class L fuse (without ARMS) results in an AFB of 92 in. and incident energy of 17.4 cal/cm^2 at 18 in. (In this case, the analysis with 85% arcing fault current required in the IEEE 1584 basic equation method, resulted in a higher AFB and incident energy than the 100% arcing fault current analysis.) When the ARMS is turned "on" by a worker, the AFB is 24.5 in. and the incident energy is 2.0 cal/cm^2. This is a significant arc flash mitigation improvement.

Figure 9-21. ARC FLASH REDUCTION MAINTENANCE SYSTEM: 20,000 A AVAILABLE SHORT-CIRCUIT CURRENT

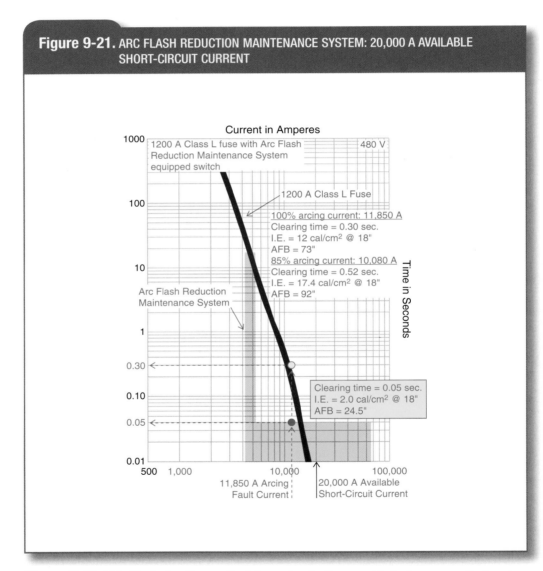

Figure 9-21. This illustrates the arc flash hazard reduction provided when an arcflash reduction maintenance system is used in conjunction with 1200A Class L fuses for a lower magnitude arcing fault current. The ARMS reduces the AFB and incident energy to lower levels than the fuse can by itself. For higher fault current levels, see Figure 9-22.

Range Where Fuse is Faster: See Figure 9-22. This scenario has 50,000 A available short-circuit current. The Class L fuses are in their current-limiting range clearing in less than a ½ cycle. Using the simplified fuse equation method, the AFB is 8 in. and incident energy is 0.39 at 18 in. The ARMS at this level with 25,930 A arcing fault current and

clearing time of 0.05 seconds results in AFB of 41 in. and incident energy of 4.7 cal/cm² at 18 in.

Example 9: Lower Fault Currents Can Result in Higher Incident Energy. As stated earlier, the IEEE 1584 basic equation method requires the AFB and incident energy to be calculated at both

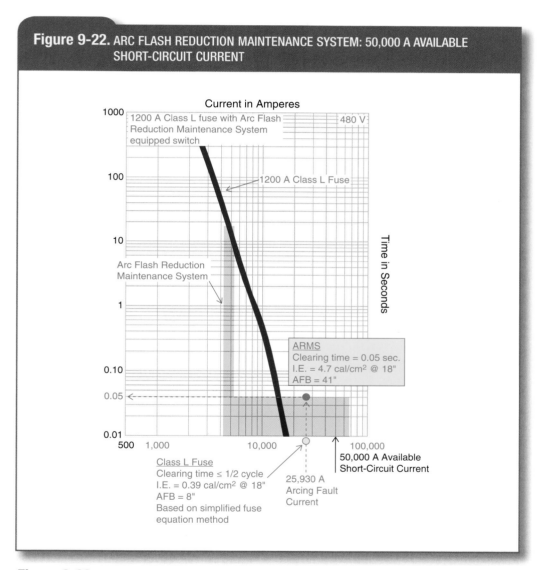

Figure 9-22. ARC FLASH REDUCTION MAINTENANCE SYSTEM: 50,000 A AVAILABLE SHORT-CIRCUIT CURRENT

Figure 9-22. This illustrates the arc flash reduction with 1200A Class L fuses and arcflash maintenance reduction system combination for higher fault current level. In this case, the Class L fuses reduce the AFB and incident energy to a much lower level than the ARMS a capable of doing at this level of fault current.

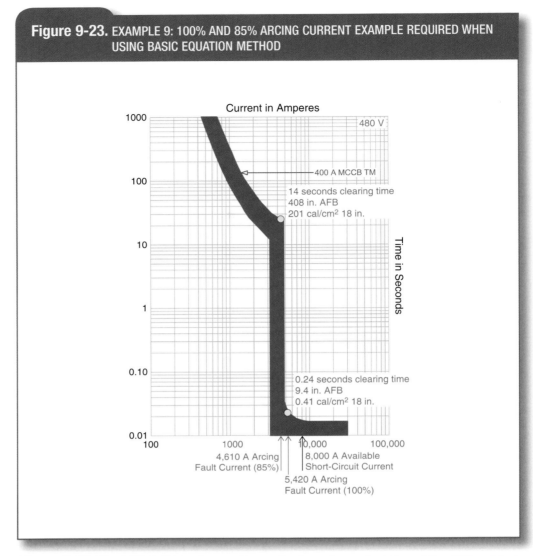

Figure 9-23. EXAMPLE 9: 100% AND 85% ARCING CURRENT EXAMPLE REQUIRED WHEN USING BASIC EQUATION METHOD

Figure 9-23. The IEEE 1584 basic equation method requires calculation of the AFB and incident energy at both 100% and 85% of the calculated arcing fault current. A small change in the arcing fault current near the instantaneous trip pickup setting can result in dramatic difference in the AFB and the incident energy.

100% and 85% of the calculated arcing fault current. **See Figure 9-23.** In this case, the 100% arcing fault current results in the circuit breaker's instantaneous trip to clear in 0.024 seconds. However, the 85% arcing fault current is below the instantaneous trip pickup and may take 14 seconds to clear. The resulting incident energy and AFB is dramatically greater for the 85% calculation. Do not assume the higher the available short-circuit currents and corresponding higher arcing fault currents mean the AFB and incident energy will be greater. It depends on the protecting-OCPD's time-current characteristics. The section covering the IEEE 1584 simplified fuse equation method later in this chapter also demonstrates this point. Many people in the electrical industry recommend high impedance transformers or installing 1:1 ratio transformers as a means to reduce the available short current believing it reduces the arc flash hazard. However, that may not be a good solution in many instances.

Figure 9-24. IEEE 1584 ARC FAULT BOUNDARY AND ARC FLASH INCIDENT ENERGY METHOD COMPARISON

2002 IEEE 1584 Method	Brief Description (Key Differentiating Points)	Comments

Simplified Fuse Equations	• Equations based on testing
	• For specific types of current-limiting fuses or fuses that open in times equal or faster than the fuse types used for IEEE 1584 testing
	• Process:
	  1. Determine the available (bolted) short-circuit current and fuse type/amp rating
	  2. Use the IEEE 1584 simplified fuse equations for the specific fuse amp rating to calculate the AFB and incident energy at 18 in. working distance (Tables and online calculator available—see Comments)
		• Suggested if the fuse type is one that is applicable to this method
		• Easy method: Eaton's Bussmann Business provides simple published tables and an online calculator
		• Fuse time–current curve is not needed
		• No need to determine the arcing current, because the equations are based on actual arcing fault tests

Figure 9-24. *IEEE 1584 arc fault boundary and arc flash incident energy method comparison. This gives a high-level overview of the three methods.*

IEEE 1584 Simplified Equations

In addition to the IEEE 1584 basic calculation method, fuse and circuit breaker "simplified equations" methods are available. These equations allow the incident energy and AFB to be calculated directly from the available 3-phase short-circuit current on a low-voltage system if the type and ampere rating of the OCPD are known. These simplified methods do not require the availability of time–current curves for the devices.

IEEE 1584 Simplified Equations: Fuses. **See Figure 9-24.** The simplified fuse equations method shown in *NFPA 70E* Informative Annex D.4.6 are from IEEE 1584. These equations are not shown in this chapter. The simplified equations are applicable for certain fuse types and for a range of available short-circuit currents. If the available short-circuit current is not within the applicable range for a specific fuse ampere rating, the simplified equations cannot be used. The IEEE

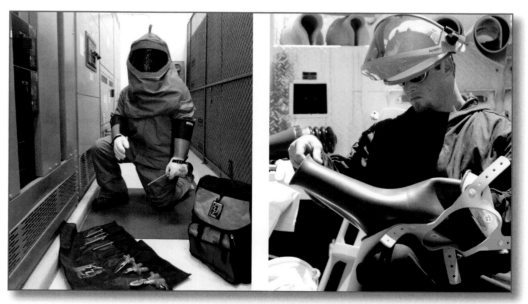

NFPA 70E general PPE provisions require, in part, that employees working within the restricted approach boundary must wear PPE in accordance with 130.4 and that employees working within the arc flash boundary must wear protective clothing and other PPE in accordance with 130.5. Courtesy of Salisbury by Honeywell.

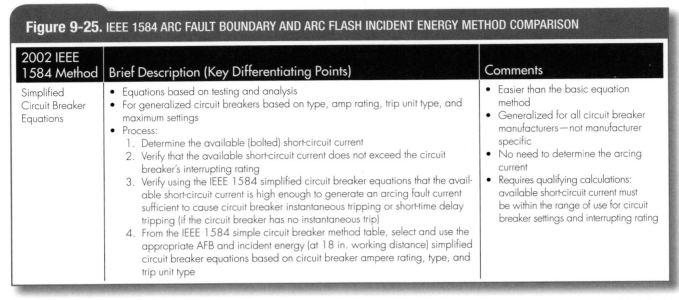

Figure 9-25. IEEE 1584 ARC FAULT BOUNDARY AND ARC FLASH INCIDENT ENERGY METHOD COMPARISON

2002 IEEE 1584 Method	Brief Description (Key Differentiating Points)	Comments
Simplified Circuit Breaker Equations	• Equations based on testing and analysis • For generalized circuit breakers based on type, amp rating, trip unit type, and maximum settings • Process: 　1. Determine the available (bolted) short-circuit current 　2. Verify that the available short-circuit current does not exceed the circuit breaker's interrupting rating 　3. Verify using the IEEE 1584 simplified circuit breaker equations that the available short-circuit current is high enough to generate an arcing fault current sufficient to cause circuit breaker instantaneous tripping or short-time delay tripping (if the circuit breaker has no instantaneous trip) 　4. From the IEEE 1584 simple circuit breaker method table, select and use the appropriate AFB and incident energy (at 18 in. working distance) simplified circuit breaker equations based on circuit breaker ampere rating, type, and trip unit type	• Easier than the basic equation method • Generalized for all circuit breaker manufacturers—not manufacturer specific • No need to determine the arcing current • Requires qualifying calculations: available short-circuit current must be within the range of use for circuit breaker settings and interrupting rating

Figure 9-25. IEEE 1584 arc fault boundary and arc flash incident energy method comparison. This gives a high-level overview of the three methods.

1584 basic equation method should not be used if the fuses are not the right type or the available short-circuit current are out of the applicable range.

***IEEE 1584 Simplified Equations: Circuit Breakers.* See Figure 9-25.** The simplified circuit breaker equation method shown in *NFPA 70E* Informative Annex D.4.7 is from IEEE 1584. If the short-circuit current falls outside the range of use for a circuit breaker setting or inter-

rupting rating, the simplified equations cannot be used. In that case, IEEE 1584 basic equations should be used. When equations are used for low-voltage power circuit breakers with short-time delay (without an instantaneous trip unit), the maximum setting (30 cycles) is assumed. Because there are several equations for the various circuit breaker types and ampere ratings, as well as the range of short-circuit currents, the equations are not reprinted here.

IEEE 1584 Resources and Tools. A variety of resources and tools have been developed that utilize the IEEE 1584 information for arc flash hazard analysis, including the materials from IEEE and from vendors.

RESOURCES AND TOOLS WITH IEEE 1584 PURCHASE. When the IEEE 1584 guide is purchased, several useful spreadsheets are provided with the actual document:
　Actual test data
　Bolted short-circuit current calculator
　Arc flash hazard calculator
　Current-limiting fuse test data
　Simplified circuit breaker method

EXAMPLE: IEEE 1584 RESOURCES AND TOOLS FROM MANUFACTURERS. The following examples demonstrate the use of

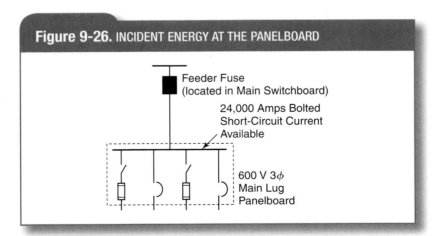

Figure 9-26. INCIDENT ENERGY AT THE PANELBOARD

Feeder Fuse (located in Main Switchboard)

24,000 Amps Bolted Short-Circuit Current Available

600 V 3ϕ Main Lug Panelboard

Figure 9-26. The incident energy at the panelboard depends on the available short-circuit current at the panelboard and the clearing time of the upstream 800-A Fuse in the main switchboard.

IEEE 1584 fuse simplified method resources and tools from manufacturers. **See Figures 9-26 and 9-27.**

Using the tables based on the IEEE 1584 simplified equation methods, the objective is to determine the incident energy for each of the following 480-volt circuits with an available short-circuit current of 24,000 amperes at the main lugs of a panelboard when protected by Low-Peak KRP-C-800SP current-limiting Class L fuses.

To use **Figure 9-27**, find the ampere rating in the header of the table. Select the available short-circuit current (bolted fault current) from the left column. Select the corresponding incident energy and AFB corresponding to the the amp rating (top horizontal column) for the available short-circuit current. For the KRP-C-800SP Fuse, the calculated incident energy is 1.46 cal/cm² with an AFB of 21 inches.

Note that if the IEEE 1584 basic equations were used, the calculated incident energy (assuming one-half cycle clearing time and a short-circuit current of 22.6 kiloamperes) would be 0.47 cal/cm² with an AFB of 10 inches for Test 3. Therefore, when the actual test data and equations for only fuses are used, the result can be higher (or in some cases lower) than the calculated value per the IEEE 1584 basic equation method.

IEEE 1584: Future Edition. The 2002 edition of IEEE 1584 represents a major contribution to furthering electrical safety in that it provided the first industry consensus standard on calculating the incident energy and AFB based on extensive arc fault testing under specific test conditions. Although the 2002 IEEE 1584 is a valuable tool, it does not provide calculation methods for all possible situations confronted in the workplace. To enhance and expand the calculation methods for arcing faults, NFPA and IEEE created a joint effort, IEEE/NFPA Collaboration on Arc Flash Research. IEEE and NFPA are working together on a research and testing initiative to enhance understanding and further mitigate the effects of hazardous electrical arc phenomena on workers, with the goal of protecting people and saving lives. Several manufacturers have donated large sums of money or equivalent services to assist this effort. The test results and information derived from these tests will be used to revise future editions of IEEE 1584 and *NFPA 70E*.

Other AFRA Calculation Methods

A recent addition in *NFPA 70E* Informative Annex D.5 focuses on DC incident energy calculations. In addition, other AC AFRA methods predate the release of the 2002 IEEE 1584; these other methods helped in the progression of quantifying arc flash hazards.

Incident Energy Analysis Method

NFPA 70E Informative Annex D.5 provides information on a direct-current incident energy analysis method. This method is provided in a paper authored by Dan Doan and was presented at the 2010 IEEE Electrical Safety Workshop. Informative

Figure 9-27. ARC FLASH INCIDENT ENERGY CALCULATOR VALUES

Fuses: Bussmann® Low-Peak® KRP-C_SP (601–1200 Amp)
Incident energy (I.E.) values are expressed in cal/cm². Arc Flash Boundary (AFB) values are expressed in inches.

Bolted Fault Current (kA)	601–800 Amp Fuse		801–1200 Amp Fuse	
	I.E.	FPB	I.E.	FPB
1	>100	>120	>100	>120
2	>100	>120	>100	>120
3	>100	>120	>100	>120
—	—	—	—	—
—	—	—	—	—
—	—	—	—	—
20	1.70	23	10.37	78
22	1.58	22	9.98	76
24	1.46	21	8.88	70
26	1.34	19	7.52	63
28	1.22	18	6.28	55
30	1.10	17	5.16	48

Full table is found in Eaton's Bussmann Business Selecting Protective Devices, Electrical Safety section.

Figure 9-27. *IEEE 1584 simplified fuse equation method put in tabular form.* Courtesy of Eaton's Bussmann Business.

Resources and Tools from Manufacturers

OTHER IEEE 1584 RESOURCES AND TOOLS ARE AVAILABLE FROM MANUFACTURERS. Eaton's Bussmann Business has developed an online Arc Flash Incident Energy Calculator to find the incident energy for low-voltage systems based on the available 3-phase short-circuit current and the type of OCPD. The calculated incident energy is valid for Low-Peak Class RK-1, Class J, and Class L fuses based on the IEEE 1584 fuse simplified equation method. This information is available using the interactive online calculator which can be accessed from the QR code shown. It is also available in tabular format in the Bussmann Selecting Protective Devices (SPD) publication. The notes for the Arc Flash Incident Energy Calculator must be read before either the online or tabular version of the calculator can be used.

For additional information, visit qr.njatcdb.org Item #1599.

Arc Flash Calculator

Notes for Arc Flash Calculator

Device Ampere Rating [400 ▼] Amps

Available 3Ø Bolted Fault Current [40] kA

[Calculate]

Send comments\questions to: fusetech@buss.com

Arc Flash Calculator

Notes for Arc Flash Calculator

Device Ampere Rating [400 ▼] Amps

Available 3Ø Bolted Fault Current [40] kA

[Calculate]

Send comments\questions to: fusetech@buss.com

Available 3Ø Bolted Fault Current: 40kA			
Overcurrent Protection Device	Incident Energy Exposure @ 18 inches (cal/cm²)	Level of PPE (cal/cm²)	Flash Protection Boundary (inches)
LPS-RK-400SP Fuse Click for the datasheet	0.25	1.2	6
400 Amp Molded Case Circuit Breaker w/ TM setting	3.08	8	34.13
Calculations generated on: 5/13/03			

Courtesy of Eaton's Bussmann Business

Annex D.5.2 provides three reference sources on DC arcing current and energy.

Ralph Lee Method: AFB and Incident Energy. The Ralph Lee AFB and incident energy calculation method, detailed in *NFPA 70E* Informative Annex D, Section D.2, is based on his seminal paper, "The Other Electrical Hazard: Electrical Arc Blast Burns," which appeared in IEEE Transactions on Industrial Applications, vol. 1A-18, no. 3, p. 246, May/June 1982. This important paper represented one of the first efforts focusing on the quantification of arcing fault hazards. The method outlined in Lee's paper is not applicable for most commercial and industrial electrical work, as these premises typically involve enclosed equipment. The incident energy from open-air arcs is less than the energy from arcs-in-a-box. Owing to this difference, it would not be suitable to determine the AFB or incident energy for an arc-in-a-box (such as arcs in electrical equipment as seen in Test 4, Test 3, and Test 1 presented in Chapter *Electrical Hazard Awareness*), because these equations would yield results lower than the expected values. The R. Lee AFB and incident energy calculation method predates IEEE 1584.

Doughty/Neal/Floyd Method: Incident Energy. The Doughty/Neal/Floyd method, detailed in *NFPA 70E* Informative Annex D, Section D.3 is based on the paper noted in Informative Annex D, Section D.3.4, by R. L. Doughty, T. E. Neal, and H. L. Floyd II. This paper, which is entitled "Predicting Incident Energy to Better Manage the Electric Arc Hazard on 600-V Power Distribution Systems," is part of the Record of the Conference Papers for the IEEE IAS 45th Annual Petroleum and Chemical Industry Conference held on September 28–30, 1998. The range of short-circuit current values, voltage, and working distances is limited compared to the IEEE 1584 method. This paper and method predate IEEE 1584.

AFRA EQUIPMENT LABELING AND DOCUMENTATION

This section will cover some aspects of arc flash risk assessment labeling and documentation. The material in this chapter augments the material on labeling and documentation in an earlier chapter, *Justification, Assessment, and Implementation of Energized Work*.

AFRA and AFRA Label Documentation

The results of the arc flash risk assessment must be documented per 130.5(A) and the data which was used to support the information that is put on the equipment label must be documented per 130.5(D) 2nd to last paragraph. 205.4 requires that OCPD "maintenance, tests, and inspections shall be documented." This documentation helps verify compliance with the 130.5(3) OCPD condition of maintenance requirement.

Equipment AFRA Label Information

130.5(D) requires the specifics of the AFRA to be put on an equipment label: nominal voltage, AFB, and results of the 130.5(C) Arc Flash PPE determination. This information must also be put on the energized electrical work permit per 130.3(B)(2)(5)a.

130.5(D)(3) permits optional ways to express the Arc Flash PPE. 130.5(D)(3)a requires the working distance to be on the label when the incident energy is marked on the label. However, the working distance is not mentioned in 130.5(D)(3)a to be included on the equipment label when the arc flash PPE category method result is marked on the label. Similarly, 130.5(D)(3)b and c do not mention including the working distance on the label for minimum arc rating of clothing or site-specific level of PPE markings. However, the working distance is provided in Table 130.7(C)(15)(A)(b) and Table 130.7(C)(15)(B) and is a variable when calculating the incident energy.

1910.335(a)(1)(i) requires that employees working in areas where there are potential electrical hazards shall be provided with, and shall use, electrical protective equipment that is appropriate for the specific parts of the body to be protected and for the work to be performed. Courtesy of Salisbury by Honeywell.

AFRA Label Information for Chapter Examples. The following is the label information for some of the arc flash risk assessment examples covered earlier in this chapter. **See Figures 9-28, 9-29, and 9-30.**

Figure 9-28. LABEL INFORMATION FOR EXAMPLE 1

Arc Flash Risk Assessment Label Information	
Nominal system voltage	480/277V
Arc flash boundary	3 feet
Arc flash PPE category	2
Working distance*	18 inches

*The working distance is not mentioned in 130.5(D)(3)a as required on the equipment label when the arc flash PPE category is used. However, the working distance is provided in Table 130.7(C)(15)(A)(b) and Table 130.7(C)(15)(B).

Figure 9-28. *Example 1.*

Figure 9-29. LABEL INFORMATION FOR EXAMPLE 4

Arc Flash Risk Assessment Label Information	
Nominal system voltage	480/277V
Arc flash boundary	8 inches
Available incident energy	0.32 cal/cm²
Working distance	18 inches

Figure 9-29. *Example 4.*

Figure 9-30. LABEL INFORMATION FOR EXAMPLE 5

Arc Flash Risk Assessment Label Information	
Nominal system voltage	480/277V
Arc flash boundary	100 inches
Minimum PPE Arc Rating	20 cal/cm²
Working distance*	18 inches

*The working distance is not mentioned in 130.5(D)(3)b as required on the equipment label when the minimum arc rating of clothing is used.

Figure 9-30. *Example 5 for low voltage power circuit breaker with short-time delay.*

Premise-Wide AFRA Labeling

Many owners conduct arc flash risk assessment for their entire premises and label the electrical equipment. Premise-wide implementation of labeling can facilitate a more efficient operation. The typical process is to update or create a complete and accurate single-line diagram of the electrical facilities, calculate the short-circuit currents, calculate the arc flash boundaries and incident energies throughout the premises, and affix AFRA labels on the electrical equipment. Premise-wide implementation of labeling can facilitate a more efficient operation. For instance, when conducting lockout/tagout, the labels readily display the arc flash risk assessment so the worker knows what is the appropriate PPE to use.

OTHER CONSIDERATIONS

There are other considerations related to arc flash risk assessments and labeling to follow.

Updating AFRA

NFPA 70E 130.5(2) requires the arc flash risk assessment to be updated when a major modification or renovation occurs. In addition, the arc flash risk assessment must be reviewed periodically, not to exceed five years, because changes in the electrical distribution systems could affect the results of the arc flash risk assessment. Per the last paragraph of 130.5, the owner of the electrical equipment is responsible for the documentation, installation, and maintenance of the field-marked labels.

Other Labeling

NEC 110.16 Arc Flash Hazard Warning Label. The *NEC* requires an arc flash hazard warning label. **See Figure 9-31.** 110.16 does not require this label to include the arc flash risk assessment information. Any electrical equipment as described in *NEC* Section 110.16 that is likely to be serviced or worked on while energized must be field or factory labeled if this equipment has been installed ac-

Figure 9-31. NEC 110.16

WARNING

Arc Flash and Shock Hazards
Appropriate PPE Required
Failure to Comply Can Result in Death or Injury
Refer to NFPA 70E

Figure 9-31. NEC 110.16 arc flash warning label.

cording to the 2002 *NEC* or more recent editions.

NEC 110.24 Marking Maximum Available Fault Current on Service Equipment. Marking the maximum available fault (bolted) current (synonymous with the maximum available short-circuit current) at the service is intended to ensure proper application of OCPD interrupting ratings and short-circuit current ratings (SCCR) of equipment. **See Figure 9-32.** When determining the maximum available short-circuit current to comply with *NEC* 110.24, methods are permitted that may result in values significantly higher

than the actual values. This outcome is acceptable as long as the equipment short-circuit current rating and interrupting ratings for the OCPDs are equal to or greater than the determined maximum available fault current.

This type of calculation is suitable for determining the estimated maximum available short-circuit current when determining an arc flash risk assessment using the arc flash PPE categories method. However, *NEC* 110.24 is not intended for incident energy analysis method calculations. Arc flash hazard incident energy method calculations should use available short-circuit current calculations that are as accurate as possible.

Arc Blast Considerations

As noted in *NFPA 70E* 130.7(A) Informational Note No. 1, physical trauma injuries may occur due to the explosive effect of some arcing fault events. The PPE requirements of 130.7 do not address protection against physical trauma other than arc flash. Little testing has been done to date on the effects and predictability of arc blast hazards. Generally, arcing fault events involving higher incident energy in short time durations result in higher arc blast energy.

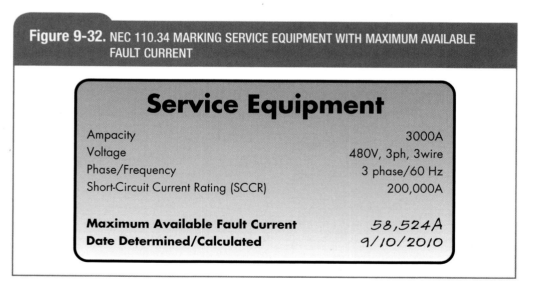

Figure 9-32. NEC 110.34 MARKING SERVICE EQUIPMENT WITH MAXIMUM AVAILABLE FAULT CURRENT

Service Equipment	
Ampacity	3000A
Voltage	480V, 3ph, 3wire
Phase/Frequency	3 phase/60 Hz
Short-Circuit Current Rating (SCCR)	200,000A
Maximum Available Fault Current	*58,524A*
Date Determined/Calculated	*9/10/2010*

Figure 9-32. Shown is 110.24 label required on service entrance equipment, with some exceptions.

Summary

This chapter provides information on the methods for determining the AFB and required arc flash PPE as part of an arc flash risk assessment. Methods are available for both AC and DC systems, and for both arcs-in-the-open and arcs-in-a-box. The arc flash PPE categories method uses tables, where conditions of use and table parameters must be satisfied before these methods can be applied. The 2002 IEEE 1584 guide provides calculation methods based on extensive testing, but the application must be within the various parameter ranges specified in IEEE 1584. For practical application of these methods, other arc flash risk assessment requirements and other considerations must be taken into account, such as the OCPD condition of maintenance, documentation, labeling, and updating of the analysis if system changes warrant.

Review Questions

1. The term "protecting-OCPD" best describes which one of the following?
   a. A defined term in Article 100, definitions, in *NFPA 70E*
   b. A term used numerous times in *NFPA 70E* but is not defined
   c. The OCPD located within the equipment being worked on
   d. The OCPD on the line side of a potential arcing fault location

2. Which of the following is/are part of the mandatory requirements in *NFPA 70E*?
   a. the arc flash PPE category method
   b. the means (formula) to determine the arc flash boundary
   c. the means (formula) to determine the available fault current
   d. the means (formula) to determine the incident energy

3. The arc flash boundary must be calculated and is the distance where the incident energy equals 1.2 cal/cm^2.
   a. True
   b. False

4. The clearing time of the "protecting-OCPD" can be longer than expected with inadequate maintenance. However, the condition of maintenance does not have to be considered when an arc flash risk assessment is performed.
   a. True
   b. False

5. Which of the following best describes what must be used to determine arc flash PPE as part of an arc flash risk assessment in accordance with *NFPA 70E*?
   a. both the arc flash PPE category method and the incident energy analysis method
   b. both the arc flash PPE category method and the incident energy analysis method with the higher value used to select the PPE
   c. either the arc flash PPE category method or the incident energy analysis method
   d. it is not necessary to determine arc flash PPE as part of an arc flash risk assessment

6. Which of the following best describes what must be included, in part, on an equipment label as part of an arc flash risk assessment in accordance with *NFPA 70E*?
   a. both the arc flash PPE category and the incident energy available
   b. neither the arc flash PPE category nor the incident energy available
   c. nominal system voltage and arc flash boundary
   d. it is not necessary to include a label on equipment as part of an arc flash risk assessment

7. The most commonly used means to calculate incident energy discussed in this chapter is __?__ .
   a. FC^2 Arcing Fault Current Calculator
   b. Point-to-point available short-circuit current method
   c. IEEE 1584
   d. Table 130.7(C)(16)

8. Which of the following best describes what must be considered to determine arc flash PPE for AC systems using the arc flash PPE categories method as part of an arc flash risk assessment in accordance with *NFPA 70E*?
   a. both the arc flash PPE category and the calculated incident energy
   b. estimated short-circuit current available, OCPD clearing time, working distance, the task, equipment condition, and whether arc flash PPE is required
   c. site-specific level of PPE, nominal voltage, and minimum arc rating of clothing
   d. it is not permitted to determine arc flash PPE using the arc flash PPE method

9. For 600 V or less systems, arcing current is typically __?__ the bolted fault current.
   a. equal to
   b. greater than
   c. less than
   d. none of the above

10. Lower fault currents can cause higher incident energy.
    a. True
    b. False

Maintenance Considerations and OCPD Work Practices

Chapter Outline

- The Value of Safety Related Electrical Maintenance
- Why is the "Condition of Maintenance" Important to the Electrical Worker?
- How a Lack of Maintenance Can Increase Hazards
- Lack or Improper Maintenance Concerns for the Electrical Worker
- Determining the Condition of Maintenance
- Frequency of Maintenance for Electrical Equipment
- Overcurrent Protective Device Work Practices

Chapter Objectives

1. Understand the value of electrical maintenance and how it can affect safety.
2. Recognize that the arc-flash hazard incident energy is related to OCPD design and the condition of maintenance.
3. Understand that electrical safety is directly tied to the condition of maintenance of other electrical equipment and the *NFPA 70E* Chapter 2 maintenance requirements.
4. Understand work practices for overcurrent protective devices (OCPDs) that can enhance worker safety.

Chapter 10

References

1. *NFPA 70E*, 2015 Edition
2. *NFPA 70B*, 2013 Edition, Recommended Practice for Electrical Equipment Maintenance
3. IEEE paper, "Prioritize Circuit Breaker and Protective Relay Maintenance Using an Arc Flash Hazard Assessment" (IEEE Paper No. ESW-2012-11)

Background

An outage was scheduled at a large semiconductor facility to conduct maintenance on the electrical distribution equipment. The cost of conducting an outage at the plant is over $1,000,000 just due to lost production. Due to these lost production costs and the disruption to production, the facility management stretched out their scheduled electrical shut-down maintenance period from three to five years. The service company that had been performing the shut-down maintenance for this facility for many years had advised the facility management that it was advisable to retain a three-year shut-down maintenance schedule. However, the facility management believed that the cooled and controlled environment where the electrical equipment is located would permit delaying the electrical maintenance two additional years.

Outage Actions

The initial step of the outage procedure is to de-energize the main power circuit breaker in each low voltage substation to reduce the load before de-energizing the main substation. To perform this step, the electrical worker wears PPE in accordance with the facility Incident Energy Analysis. The Electrical Worker actuated the "trip" button on the door of the main power circuit breaker and nothing happened. The power remained on, the circuit breaker flag did not change state and the discharge of the stored energy mechanism was not heard. The Electrical Worker stepped out of the room and reviewed as-built drawings of the switchgear with the maintenance service company representative and the facility engineer. While they were reviewing the drawings looking for any additional interlocks that may be blocking the "trip" signal to the circuit breaker, the power finally went off and the discharge of the circuit breaker stored energy mechanism was heard, indicating the circuit breaker finally opened.

Outage Findings

The main power circuit breaker was disassembled and inspected during the outage. The main latch release mechanism lubrication was hardened and unmalleable. When the initial trip signal was received, the spring mechanism was unable to apply enough force to open the circuit breaker due to the hardening of the lubrication. Over a few minutes, the spring eventually overcame the hardened lubrication, thereby opening the circuit breaker.

Conclusions

The power circuit breakers instruction manual recommended "exercising" the circuit breaker every six months. The purpose of this recommendation is to keep the lubrication in the operating mechanism malleable. If a fault had occurred on the load side of the circuit breaker, the circuit breaker would not have operated as anticipated, resulting in a higher than calculated clearing time. The resulting incident energy would have been greater than 42 calories/cm^2 before the upstream fuse on the primary of the transformer would have cleared the fault. The semiconductor company modified their maintenance practices to incorporate regular "exercising" of their circuit breakers and reduced their outage interval back to 3 years.

Source: Courtesy of Eaton Corporation

INTRODUCTION

Maintenance of electrical power systems is required by *NFPA 70E* and it affects the safety of the Electrical Worker. Specific discussions are presented on the effects of improperly operating overcurrent protective devices (OCPDs) and how they might affect the determination of the incident energy or arc flash PPE category when conducting an Arc Flash Risk Assessment. Examples of real world maintenance issues are demonstrated, focusing on how they have or might have resulted in safety issues for the Electrical Worker. The topic of determining the condition of maintenance of OCPDs is discussed to allow the Electrical Worker to evaluate the risks involved in deciding if a specific OCPD should be used in the Arc Flash Risk Assessment. The frequency of maintenance of specific OCPD is covered to assist the Electrical Worker with providing proper maintenance for electrical power systems. In addition, maintenance for electrical equipment, in general, is now a *NFPA 70E* requirement.

Proper OCPD device work practices are important to worker safety. Electrical Workers should be trained in and use safe work practices regarding OCPDs. If a fault has resulted in opening an OCPD, OSHA and *NFPA 70E* do not permit resetting a circuit breaker or replacing fuses until it is safe to do so. After a circuit breaker interrupts a fault, testing the circuit breaker may be required. There are means to reduce the arc-flash hazard when racking circuit breakers. Fuses should be tested properly.

THE VALUE OF ELECTRICAL SAFETY-RELATED MAINTENANCE

NFPA 70E continues to evolve, and the 2015 Edition elevated the importance of electrical equipment maintenance for electrical safety-related work practices. The first sentence of 90.2, which is the Scope for *NFPA 70E*, is expanded to include "This standard addresses . . . safety-related maintenance requirements . . ."

The evolving awareness that electrical equipment condition of maintenance is directly related to electrical safe work practices is evident by the new 110.1(B) requirement. This new 2015 *NFPA 70E*

requirement mandates the employer's electrical safety program to include consideration for condition of maintenance of electrical equipment and systems. A well-administered maintenance program will minimize costly breakdowns and unplanned shutdown of production equipment or loads. Equipment repair cost and equipment downtime records can document direct, measurable, economic benefits after a continuing electrical maintenance program has been properly implemented. The electrical industry *and NFPA 70E* now emphasizes that maintaining electrical equipment is an important element for electrical safe work practices.

The purpose of maintenance requirements in *NFPA 70E* is stated in 200.1(3) as preserving or restoring the condition of electrical equipment for the safety of employees who are exposed to electrical hazards. Performing regular maintenance of electrical equipment is proven through IEEE studies to significantly reduce electrical failure rates and results in increased reliability, reduction in repair costs, while allowing efforts to be concentrated on improving operation rather than responding to emergency repairs. Electrical maintenance is essential to reduce hazards to Electrical Workers and property that can result from electrical equipment malfunction or failure. Although electrical equipment deterioration is normal, equipment failure is not inevitable. In addition to normal deterioration, there are other potential causes of equipment failure that can be detected and corrected through a continuing electrical maintenance program. Among these are load changes or additions, circuit alterations, improperly set or improperly selected protective devices, and changing voltage conditions.

OSHA Tip

1910.303(g) 600 Volts, nominal, or less. This paragraph applies to electric equipment operating at 600, volts, nominal, or less to ground. 1910.303(g)(1) Space about electric equipment. Sufficient access and working space shall be provided and maintained about all electric equipment to permit ready and safe operation and maintenance of such equipment.

NFPA 70E Maintenance Requirements for OCPDs

205.4: requires OCPDs to be maintained per manufacturers' instructions or industry consensus standards. "Maintenance, tests, and inspections shall be documented."

210.5: requires OCPDs to be maintained to safely withstand or be able to interrupt the available fault current. Informational Note makes mention that improper or lack of maintenance can increase arc flash hazard incident energy.

225.1: requires fuse body and fuse mounting means to be maintained. Mountings for current-limiting fuses cannot be altered to allow for insertion of non-current-limiting fuses.

225.2: requires molded cases circuit breaker cases and handles to be maintained properly.

225.3: requires inspection and testing circuit breakers that interrupt fault current approaching their interrupting rating.

NFPA 70E Maintenance of Other Equipment

In addition to maintenance requirements for OCPDs, *NFPA 70E* contains requirements for other parts of the electrical power system.

205.3: requires electrical equipment to be maintained per manufacturers' instructions or industry consensus standards. The owner or his representative is responsible for maintenance and documentation.

The following are Articles of *NFPA 70E* Chapter 2 which have maintenance requirements for various types of equipment:

200 Introduction
205 General Maintenance Requirements
210 Substations, Switchgear Assemblies, Switchboards, Panelboards, Motor Control Centers (MCC), and Disconnect Switches
215 Premises Wiring
220 Controller Equipment
225 Fuses and Circuit Breakers
230 Rotating Equipment
235 Hazardous (Classified) Locations
240 Batteries and Battery Rooms
245 Portable Tools and Equipment

WHY IS THE "CONDITION OF MAINTENANCE" IMPORTANT TO THE ELECTRICAL WORKER?

Many workers, managers, and owners assume that once installed, electrical equipment will work as originally designed for the life of the system. Lack of proper maintenance introduces multiple modes of potential failure into the electrical system that cannot be seen until problems are proactively eliminated through maintenance—or until there is a system failure.

Not performing maintenance on an electrical system can have similar consequences to not performing maintenance on your car. You may skip checking your brakes, regularly inflating your tires, or changing your oil, and your car seems to run fine. But if you do not perform regular maintenance on your car, you risk not being able to brake in time to avoid an accident, your tires may fail due to lack of pressure, causing your car to lose control, or your engine could burn up. The mechanical and insulation systems in an electrical system also need periodic maintenance to perform in accordance with their design.

So why is the condition of maintenance important to the Electrical Worker? Improper or lack of maintenance on an overcurrent protective device can lead to higher arc flash incident energy than may have been determined by an arc flash risk assessment. If the incident energy for an actual arc flash event is higher than the incident energy calculated by the arc flash risk assessment, the Electrical Worker may not be wearing sufficient PPE. In addition, workers who thought they were outside the arc flash boundary may be wearing flammable or meltable garments and subject to clothing ignition and other unanticipated consequences. These are among the reasons why *NFPA 70E* Chapter 2, "Safety-Related Maintenance Requirements," specifically requires overcurrent protective devices to be maintained properly and documented (205.4). Similarly, when a worker has to remove the trim from an energized electrical panel to perform trouble-shooting diagnostics, that worker may be at greater risk of injury or death if the internal wiring, terminations, devices, etc. have not been properly maintained.

The importance of "properly maintained" as a condition of safe work practices is stated in many *NFPA 70E* requirements and Informational Notes: see Arc Flash Hazard definition Informational Note No. 1, 130.2(A)(4), 130.2(A)(4) Informational Note, Table 130.7(C)(15)(A)(a) column heading "Equipment Condition*" (including * note at end of table), and 130.5(3) (including 130.5 Informational Notes No. 1 and No. 4). It stands to reason a fundamental principle of safe work practices requires electrical equipment to be in good condition and properly maintained.

To investigate the consequences of OCPDs not operating when needed for arc flash mitigation, a large industrial conducted a simulation and published a technical paper. In this simulation, the first level OCPDs supplying the equipment (arc flash protecting OCPDs) were turned "off" in the commercial arc flash analysis software package as if the OCPDs would not operate. In this state, the arc flash mitigation then was relegated to the next higher level upstream OCPD, which is a larger ampere rated device. The IEEE paper, "Prioritize Circuit Breaker and Protective Relay Maintenance Using an Arc Flash Hazard Assessment" (IEEE Paper No. ESW-2012-11) by Dan Doan, examines the effects on an Arc Flash Hazard Assessment (now Arc Flash Risk Assessment in *NFPA 70E* 2015) in an industrial facility, if the upstream protective devices do not operate correctly. The study determined that if the closest upstream circuit breaker or relay did not operate and the next higher level upstream OCPD cleared the arcing fault current, approximately in 2/3 of the cases, the worker would not have adequately arc rated PPE.

HOW A LACK OF MAINTENANCE CAN INCREASE HAZARDS

As part of the Arc Flash Risk Assessment, as required in *NFPA 70E* 130.5, the arc flash boundary (AFB) and PPE need to be determined for Electrical Workers to perform work within the AFB. The design, opening time, and condition of maintenance of the overcurrent protective devices must be taken into account as part of this assessment per 130.5(3). **See Figure 10-1.** This table shows how the magnitude of the arcing fault current and/or overcurrent protective device clearing time will affect incident energy when the working distance is held constant at 18 inches from the arcing fault. Note: The incident energy calculations were performed using IEEE 1584 *Guide for Performing Arc-Flash Hazard Calculations,* basics equation method for "Arc in a Box".

Looking at the table, if the available short-circuit current is 22KA, an upstream protective device that could operate and clear the fault in 0.05 seconds would result in an incident energy of 2.11 cal/cm^2. In comparison, if the upstream protective device required 1 second to clear the fault, the resulting incident energy is 42.14 cal/cm^2. It is clearly evident that the "Clearing Time" of the protective device has a major influence on the resulting incident energy during an arcing fault event. If the protective device fails to operate as designed, or not at all, the incident energy that the Electrical Worker will be exposed to will be much higher than calculated. This issue is why *NFPA 70E* requires "the condition of maintenance" of the arc flash protecting OCPD to be taken into consideration when performing an Arc Flash Risk Assessment.

Figure 10-1. INCIDENT ENERGY FOR VARIOUS SCENARIOS

Distance (inches)	18	18	18	18	18	18	18	18
Clearing Time (cyc)	3	6	10	20	30	40	50	60
(sec)	0.05	0.10	0.17	0.33	0.50	0.67	0.83	1.00
SC kA	cal/cm^2	cal/cm^2	cal/cm^2	cal/cm^2	cal/cm^2	cal/cm^2	cal/cm^2	cal/cm^2
16	2.07	4.14	6.90	13.81	20.71	27.62	34.52	41.42
18	2.03	4.06	6.76	13.52	20.29	27.05	33.81	40.57
20	2.04	4.08	6.80	13.60	20.40	27.21	34.01	40.81
22	2.11	4.21	7.02	14.05	21.07	28.09	35.11	42.14
24	2.23	4.46	7.43	14.85	22.28	29.71	37.13	44.56
26	2.40	4.81	8.01	16.02	24.04	32.05	40.06	48.07
28	2.63	5.27	8.78	17.56	26.34	35.12	43.90	52.67
30	2.60	5.84	9.73	19.46	29.18	38.91	48.64	58.37
32	3.26	6.52	10.86	21.71	32.58	43.44	54.30	65.16
34	3.65	7.30	12.17	24.35	36.52	48.69	60.86	73.04
36	4.10	8.20	13.67	27.34	41.00	54.67	68.34	82.01
38	4.60	9.21	15.34	30.69	46.03	61.38	76.72	92.07
40	5.16	10.32	17.20	34.41	51.61	68.81	86.02	103.22
42	5.77	11.55	19.24	38.49	57.73	76.96	96.22	115.46
44	6.44	12.88	21.47	42.93	64.40	85.87	107.33	128.80
46	7.16	14.32	23.87	47.74	71.61	95.49	119.36	143.23
48	7.94	15.87	26.46	52.92	79.37	105.83	132.29	158.75
50	8.77	17.54	29.23	58.45	87.68	116.91	146.13	175.36

Figure 10-1. *This table provides the incident energy for various available short-circuit currents (left vertical axis) and various OCPD clearing times (horizontal axis).*

LACK OR IMPROPER MAINTENANCE CONCERNS FOR THE ELECTRICAL WORKER

Electrical power distribution equipment that is not properly maintained can create a dangerous situation for the Electrical Worker. Maintenance is not just limited to overcurrent protective devices, as un-maintained bus systems, conductors, conductor terminations, insulators, and enclosures can also create hazards. The following are examples from one of Eaton's reconditioning facilities that consistently services circuit breakers, contactors, starters, and switches that suffer from a lack of lubrication.

Example 1

See Figure 10-2. This circuit breaker is less than ten years old but the lubrication had dried out and was contaminated by dirt. The lack of lubrication caused the shutter mechanism to freeze, broke a spring, and rendered the circuit breaker inoperable.

Many equipment owners do not realize that the operating mechanisms of circuit breakers and many switches are lubricated when they are built and that the lubrication dries out over time, even if the breaker (or switch) has not been operated. Just because the device does not operate does not mean that it does not need to be maintained.

Figure 10-2. CONTAMINATED LUBRICATION

Figure 10-2. The circuit breaker lubrication is dried out and contaminated. This caused the shutter interlock to stop working, broke a spring, and rendered the breaker inoperable. *Courtesy of Eaton Corporation.*

Example 2

An Electrical Worker may be misled into the "state" of the breaker by lack of maintenance. In **Figure 10-3,** the dry unlubricated mechanism caused the "Open / Close" indicator to be stuck in the "Open Condition" even though the circuit breaker is closed. This issue creates an extremely unsafe condition, as the electrical worker may be misled into thinking a circuit breaker is "open" when it is really "closed". If the Electrical Worker were to try to rack-in or rack-out the circuit breaker while closed, they may be relying on safety interlocks to open the breaker prior to connecting or disconnecting from the stabs.

Example 3

Not only is periodic maintenance necessary, the maintenance needs to be performed by workers qualified for the task. The following is another example of improper maintenance creating a safety situation for the Electrical Worker. In

Figure 10-3. STUCK INDICATOR

Figure 10-3. Mechanism is completely dry causing "Open / Close" indicator to be stuck in the "Open Condition" even though the breaker is closed. *Courtesy of Eaton Corporation.*

Figure 10-4. IMPROPER LUBRICATION

Figure 10-4. Improper lubrication of primary stab finger clusters leads to overheating and severe equipment damage. Courtesy of Eaton Corporation.

Figure 10-4, the improper lubricant was applied to the finger clusters that make contact with the cell stabs for a draw-out circuit breaker mechanism. The improper type lubricant caused the finger clusters to overheat, leading to a catastrophic failure of the circuit breaker and cell structure. Improperly maintained equipment like this can be a source of an arc flash incident.

Example 4

Abnormal environmental conditions require more maintenance than climate controlled environments. For example, "silver whiskers" (**see Figure 10-5**) have been found to grow on sharp edges of silver coated parts in sulfur rich, warm, moist environments. These whiskers have been the cause of many unexpected electrical equipment failures in the petrochemical and food industry.

Example 5

Verifying the proper operation of electrical equipment components is an important part of performing electrical equipment maintenance. Most outdoor electrical installations use "space heaters" to maintain temperature and low moisture levels inside electrical equipment enclosures. If the space heaters are inoperative, moisture can condense on the equipment bus structures and conductor terminations. trapping dirt which leads to tracking and eventual insulation failure. **See Figures 10-6 and 10-7.** Unmaintained equipment like this can be a serious safety concern when workers remove equipment covers or open doors. The vibrations from the workers' actions may be the final action

Figure 10-5. SILVER WHISKERS

Figure 10-5. "Silver Whiskers" caused by a contaminated environment and lack of maintenance led to control power transformer failure on this medium voltage switchgear. Courtesy of Eaton Corporation.

Figure 10-6. HEATER FAILURE

Figure 10-6. Heater failure leads to moisture build-up and bus failure. Regular maintenance would have detected the defective heater. Courtesy of Eaton Corporation.

Figure 10-7. BUS FAILURE

Figure 10-7. Lack of maintenance can lead to bus failure as dirt, moisture and contaminates cause tracking. Courtesy of Eaton Corporation.

Figure 10-8. WELDED CONTACT

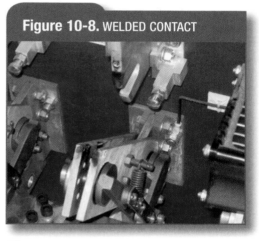

Figure 10-8. Welded contact pole on bolted pressure switch from lack of maintenance and exercising. Electrical Worker did not verify switch was completely open and as a result the worker was injured due to shock. Courtesy of Eaton Corporation.

whereby the deteriorating insulation results in a high energy arcing fault.

Example 6

Some electrical equipment, if it is not exercised regularly, can seize and contacts weld shut. In the following example a bolted pressure switch was never exercised since its original installation. While the operating mechanism worked properly, one of the moving contacts welded to a stationary contact due to lack of maintenance. **See Figure 10-8.** The electrical worker opened the switch and assumed all three phases of the switch opened since he heard the mechanism operate. Unfortunately, the worker did not perform a voltage test to verify the switch was open. The worker was injured due to shock. The worker should have been wearing shock and arc flash PPE and should have followed proper work practices to verify the switch was in an electrically safe work condition.

DETERMINING THE CONDITION OF MAINTENANCE

As part of the arc flash risk assessment, *NFPA 70E* 130.5(3) requires that that "condition of maintenance" must be considered for the overcurrent protection device used in the assessment determination. So how does the Electrical Worker

Background

The Hazard Associated with Misapplication of an Overcurrent Protective Device—Power Fuse or Circuit Breakers

Most misapplication problems are avoided when the design professional, equipment suppliers and equipment installers comply with the installation and design requirements of the *National Electrical Code* (*NFPA 70*). An Electrical Worker will often be working at a facility which is unfamiliar and unable to recognize only the most obvious misapplications of power system components. This will give an Electrical Worker a reason to use PPE even when simply working near the equipment, rather than working on the equipment.

The most frequent misapplication of an overcurrent protective devices is the use of power fuses or circuit breakers that have an interrupting rating that is less than the available short-circuit current that power sources can deliver.

Courtesy of Eaton Corporation

Example: The equipment user retrofitted a surge-protection device into an existing switchboard assembly but used a molded-case circuit breaker that had an inadequate interrupting-current rating. The surge-protection device failed, causing a short-circuit. The short-circuit developed into an arcing fault. Nobody was interacting with the equipment, but if a person had been near the equipment when this arcing fault occurred, that person would likely have been injured if not wearing proper PPE.

Assemblies such as switchboards, motor-control centers or switchgear have associated ratings called Short-Circuit Current, Short-Circuit Withstand, Momentary Current or Short-Time Current Ratings. The rating of the assembly is typically limited to the lowest rated over current protective device in the assembly. For example if a switchboard assembly has (5) 65KA interrupting rated circuit breakers and (1) 35KA interrupting rated circuit breaker, the switchboard assembly is considered to only have a 35KA short-circuit current rating even if the switchboard assembly bus short-circuit current rating is good for 65KA. A catastrophic failure might put the Electrical Worker at risk if a circuit breaker or fuse attempts to interrupt a fault current greater than its interrupting rating.

determine if the overcurrent protective device is properly maintained?

When electrical equipment is installed, it is advisable to conduct site acceptance tests to verify that the electrical equipment is performing in accordance with the design professional's plans and specifications. These tests also verify that components, like overcurrent protective devices, are performing within manufacturer's tolerances and industry standards while providing a baseline for future maintenance testing. This service would also include setting the overcurrent protective devices

per a coordination study and possibly an incident energy analysis to provide the electrical system the desired level of coordination while minimizing incident energy levels. If these settings are not properly implemented, the Electrical Worker may be exposed to a higher incident energy level than anticipated. After these tests are completed, a service sticker is placed on the devices tested to indicate the date of service (condition of maintenance confirmation). **See Figure 10-9**. An Electrical Worker can use this data, and the corresponding service report, to deter-

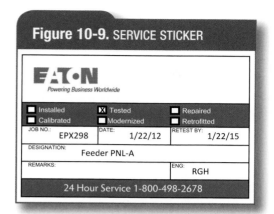

Figure 10-9. SERVICE STICKER

Figure 10-9. *A typical service sticker that is affixed to electrical equipment that indicates the date of service and the retest date which would assist in determining the condition of maintenance. Courtesy of Eaton Corporation.*

mine if the equipment has been tested and inspected within a recommended frequency of maintenance for the type of equipment.

Alternative to OCPDs that Are Not Maintained Properly

So what if the equipment has never been serviced or documentation is not available? Unfortunately this is fairly common occurrence in many electrical system installations. At a minimum, the following inspections should be performed to determine the risk that the equipment may not function as intended.

1. Is the environment in which the equipment is located isolated or protected from the elements and contaminates (dust and chemicals)? Clean and dry equipment is less likely to experience deterioration with insulation and contamination to lubricants used in overcurrent protective devices.

2. Are the overcurrent protective devices in visibly good condition? Circuit breakers may become discolored if subjected to overheating conditions. Also, cracks or burn marks might be visible, indicating that the OCPD has been subjected to an event that may affect the performance of the OCPD.

3. What is the application of the equipment and how often is it operated? Some applications may put more stress on equipment, including thermal loading which could lead to deterioration of insulation. While an extreme number of operations might wear out mechanical systems of circuit breakers eventually, some operations provide an indication that the mechanical systems are working and lubricated.

4. What is the age of the equipment? Older equipment, especially if it has not been maintained or operated, is a higher risk to lack of operation than newer equipment. Newer lubricants used in some circuit breaker brands are more durable and less likely to harden and prevent proper operation.

Even if the environment is controlled and very clean, there isn't any way to know if the circuit breaker lubrication has hardened and may prevent the device from operating within its operating parameters.

NFPA 70E 130.5(3) states to "take into consideration" the OCPD condition of maintenance but *NFPA 70E* does not provide specific requirements or actions to take if the arc flash protecting OCPD is maintained inadequately or improperly. There is no single answer to this situation and the action taken may depend on the circumstances of a specific situation. It can be factored into a risk assessment procedure; see 110.1(G), Risk Assessment Procedure, and Informative Annex F, Risk Assessment Procedure. If the original objective was to perform justified energized work, it may be advisable to change the work plan and establish an electrically safe work condition rather than work on the equipment while energized. However, the Electrical Worker still has to interface with the equipment to check for the absence of voltage in the process of putting the equipment in an electrically safe work condition. If the task is to open a disconnect or rack-out a circuit breaker, using a remote operator or remote racking device may be a more appropriate approach.

Again, checking for the absence of voltage still may have to be performed. A maintenance manager for a large automotive industrial has a policy to investigate upstream OCPDs to locate an OCPD device which has a suitable condition of maintenance and use that device for the arc flash risk assessment. In many cases, industrials perform periodic maintenance on the larger OCPDs in their systems. With this approach, the AFB and AF PPE determined will have greater values than if the nearest upstream OCPD is used for the assessment.

FREQUENCY OF MAINTENANCE FOR ELECTRICAL EQUIPMENT

Since maintenance of electrical equipment (other than OCPDs) consists mainly of keeping it clean and properly tight conductor terminations, the frequency of scheduled inspection and maintenance depends strongly on the cleanliness of the surroundings. Cleaning and preventive measures are a part of any good maintenance program. Facility operating and local conditions can vary to such an extent that the actual schedule should be tailored to the environment the equipment is located in. When the equipment is subject to a clean and dry environment, cleaning is not required as frequently as when the environment is humid with a significant amount of dust and other foreign matter. It is recommended that maintenance record sheets be completed for the equipment. Careful and accurate documentation of all maintenance activi-

Background

Courtesy of Eaton Corporation

Two Electrical Workers were injured as they removed the dead front from a panel that was energized. An arcing fault occurred, with both employees sustaining burns to their hands, and one of the employees receiving burns to his face. They were not wearing PPE. The employees were attempting to identify and relocate a lighting circuit to allow some other crafts to continue to work in the area. It was also discovered that the EMT connector coming into the bottom of the panel did not have a bushing. The conductors had been rubbing on the junction point of the EMT to the panel for an extended time, and when the dead front was being moved the insulation on the conductors finally wore through making contact with the panel or EMT. Do not assume that smaller lighting panels do not have the potential to create a dangerous arcing fault condition. Always verify that equipment has been installed and is maintained regardless of the equipment size and rating and wear proper PPE per an arc flash risk assessment.

ties provides a valuable historical reference on the electrical equipment's condition of maintenance over time.

Frequency of Maintenance for Overcurrent Protective Devices

The frequency of maintenance and maintenance recommended can vary by the OCPD type and technologies.

Power Circuit Breakers. The industry standard for power circuit breakers recommends a general inspection and lubrication after the number of operations listed in **Figure 10-10**. The inspection should also be conducted at the end of the first six months of service, even if the number of operations has not been reached.

Almost all power circuit breakers require periodic renewal of lubrication in their operating mechanism. There are four factors that determine the frequency that lubrication should be renewed:

1. The continuous current rating of the circuit breaker
2. The number of close-open operations since the most recent renewal
3. The time since the most recent renewal
4. The circuit breaker's operational environment

Manufacturing Standard ANSI/IEEE C37.16 establishes endurance requirements for low-voltage power circuit breakers. These requirements relate the minimum number of close-open operations that a breaker must be able to accomplish before requiring service. One of the limiting factors is the need to renew lubrication in a circuit breaker's mechanism. In general, the greater the continuous current rating of the circuit breaker, the fewer operations before required service. The manufacturer's instructional literature is allowed to suggest a greater number of operations than the number given in the manufacturing standard. For example, although an 800 ampere-rated circuit breaker is required by the manufacturing standard to endure 500 operations before service is needed, a manufacturer's instruction book might indicate that an 800 ampere

Figure 10-10. POWER CIRCUIT BREAKERS

Breaker Frame Size	Interval[1] (Breaker Cycles)
800 amperes and below	1750
Between 800 and 3000 amperes	500
3000 amperes and above	250

[1]Breaker Cycle = one no load open/close operation

Figure 10-10. Industry standard for power circuit breakers recommends a general inspection and lubrication after specified number of operations.

would require renewal of lubrication after 1750 operations.

The first maintenance inspection should be conducted after the first six months of service to determine the cleanliness of the environment and the number of operations. Subsequent checks should be made at least once a year and may be extended longer if these maintenance inspections indicate that there are no problems. If the maintenance inspection shows a heavy accumulation of dirt, lint or foreign matter, the service interval should be decreased. Any circuit breaker that has interrupted a fault should be serviced prior to being placed back into operation.

Molded Case Circuit Breakers. Molded case circuit breakers, both thermal magnetic trip and electronic trip types, have proven to be very reliable in service. They inherently require very little maintenance due to their enclosed design. A molded case circuit breaker must be operated open and closed with sufficient frequency to ensure that its main contacts are cleaned by wiping action and that the lubrication materials within its mechanism remain evenly spread. For any circuit breaker that is not operated in its normal service, a periodic open-close exercise should be planned. It is recommended that the circuit breaker mechanism be exercised annually. Some studies have shown that the absence of annual

OSHA 1910.334(b)(2):
Reclosing circuits after protective device operation. After a circuit is de-energized by a circuit protective device, the circuit may not be manually reenergized until it has been determined that the equipment and circuit can be safely energized. The repetitive manual reclosing of circuit breakers or re-energizing circuits through replaced fuses is prohibited.

Note: When it can be determined from the design of the circuit and the over current devices involved that the automatic operation of a device was caused by an overload rather than a fault condition, no examination of the circuit or connected equipment is needed before the circuit is reenergized.

mechanical cycling has not impacted reliability possibly due to advanced Teflon based lubricants applied in newer designs by some manufacturers. Inspection and maintenance may be required more frequently if adverse operating or environmental conditions exist.

There isn't a universal recommendation on whether to test or not to test molded case circuit breakers by primary injection at any regular interval. When testing thermal-magnetic molded case circuit breakers in the field, specific ambient conditions and cool down periods cannot be simulated to produce results that would be comparative to established UL and IEC performance standards. NEMA AB-4 "Guidelines for Inspection and Preventive Maintenance of Molded Case Circuit Breakers Used in Commercial and Industrial Applications" provides guidance on testing of a molded case circuit breaker in the field taking into account these limitations.

Electronic trip units were developed to resolve the difficulties in field testing of molded case circuit breakers. The use of an electronic trip unit allows the circuit breaker to be tested by secondary current injection which simulates the fault current to the trip unit at the secondary of the circuit breaker current transformers. This test verifies the trip

actuator, latch mechanism and electronic circuit of the circuit breaker. Other advantages of secondary injection test sets is that they are lightweight, simple to operate, and have low power requirements making testing easier and requiring less downtime.

Fuses. Fuses lack moving exterior parts so physical maintenance is very minimal. However, periodically checking fuse bodies, fuse mountings, and adjacent conductor terminations for signs of overheating, poor connections, or insufficient conductor ampacity is important. Fuses cannot have their performance characteristics tested, but most manufacturers will provide a micro-ohm or millivolt drop test reading that indicates that the fuse would operate within specifications.

Fuses are typically used in conjunction with disconnects which require periodic inspection and maintenance. If a disconnect has lubricated mechanisms, then it is necessary to maintain the lubrication properly. In applications where stored energy type disconnects are equipped with a ground-fault protection relay, the disconnects should be periodically inspected and maintained, and the ground-fault relay calibrated. After a stored energy type disconnect interrupts a ground fault, the disconnect and mechanism should be inspected and maintained if necessary.

Like all electrical equipment, the frequency of maintenance is highly dependent upon the environment the equipment is subject to. An inspection during the first year of service is recommended to determine the frequency of maintenance based upon observed accumulation of dirt or foreign matter. De-energized maintenance and cleaning should not exceed 3 years per the recommendations of *NFPA 70B* 2013. The manufacturer's instruction book should be consulted as the design of the equipment may require more frequent maintenance.

NFPA 70B Frequency of Maintenance

NFPA 70B, Recommended Practice for Electrical Equipment Maintenance, provides some frequency of maintenance guidelines as well as guidelines for setting up an electrical preventive maintenance (EPM) program, including sample forms and requirements for electrical system maintenance. All users of electrical equipment are encouraged to utilize an electrical preventative maintenance and testing program designed for their particular application. The methods of establishing such a program is covered in Section 6 of *NFPA 70B* and Annex K "Long Term Maintenance Guidelines" has recommendations in a table form for various types of equipment showing tests and frequencies should manufacturer data not be available.

NFPA 70B recommends that once the initial frequency for inspection and tests has been established, this frequency should be adhered to for at least four maintenance cycles unless unexpected failures occur. If unexpected failures occur, the interval between inspections should be reduced by 50% as soon as the trouble occurs. After four cycles of inspections have been completed, a pattern should have developed, and if the equipment does not require service, the inspection period can be extended by 50%. The adjustment of the interval between inspection and testing should be adjusted for the optimum interval.

OVERCURRENT PROTECTIVE DEVICE WORK PRACTICES

Electrical Workers deal with circuit breakers and fuses on a regular basis, from installing these devices on new installations, to performing diagnostics, to replacing them. There are several important safe work practices that should be implemented when working with circuit breakers and fuses, including maintenance and diagnostic tips.

Resetting Circuit Breakers or Replacing Fuses

A circuit breaker should not be reset, nor should fuses be replaced, until the cause of the problem is known and rectified and it has been verified that it is safe to reenergize the circuit.

NFPA 70E 130.6(M) is essentially the same requirement with slightly different wording.

This work practice is important for the safety of the workers. If an OCPD opens as a result of fault conditions, damage at the point of the fault could result. If the source of the fault is not located and corrected, reclosing the OCPD into an uncleared fault might result in an even more severe fault. If the protective device is a circuit breaker, it could have been damaged during the initial interruption (see *NFPA 70E* 225.3). For this reason, following proper procedures after an OCPD has interrupted a fault is important. If the cause of a circuit breaker or fuse opening is an overload, then it is permissible to merely reset the circuit breaker or replace the fuses.

Circuit Breakers

Important work practice considerations should be implemented when an electrical installation includes circuit breakers. These include what to do after a circuit breaker interrupts a fault, racking circuit breakers, or replacing or adding circuit breakers to a panel.

Circuit Breaker Evaluation After Fault Interruption. After fault interruption, it is necessary to evaluate a circuit breaker for suitability of use before placing it back into service. This evaluation requires a thorough inspection and may require electrical testing to specifications according to the manufacturer's instruction manual. If the circuit breaker manufacturer's procedures are not available, it is advisable to use an industry standard such as NEMA AB-4. *NFPA 70E* 225.3 requires inspection and testing a circuit breaker per manufacturer's instructions after a circuit breaker has interrupted a fault current approaching its interrupting rating.

The following is an insightful passage written by Vince A. Baclawski, Technical Director, Power Distribution Products,

National Electrical Manufacturers Association (NEMA) in Electrical Construction & Maintenance Magazine, January 1995, p. 10. Copyright EC&M (January 1995). Reprinted by permission of Penton media.

After a high level fault has occurred in equipment that is properly rated and installed, it is not always clear to investigating electricians what damage has occurred inside encased equipment. The circuit breaker may well appear virtually clean while its internal condition is unknown. For such situations, the NEMA AB4 "Guidelines for Inspection and Preventive Maintenance of MCCBs Used in Commercial and Industrial Applications" may be of help. Circuit breakers unsuitable for continued service may be identified by simple inspection under these guidelines. Testing outlined in the document is another and more definite step that will help to identify circuit breakers that are not suitable for continued service.

After the occurrence of a short circuit, it is important that the cause be investigated and repaired and that the condition of the installed equipment be investigated. A circuit breaker may require replacement just as any other switching device, wiring or electrical equipment in the circuit that has been exposed to a short circuit. Questionable circuit breakers must be replaced for continued, dependable circuit protection.

Figure 10-11. REMOTE RACKING

Figure 10-11. *A remote racking device permits a worker to rack circuit breakers in or out at a safe distance from the equipment.* Courtesy of Eaton Corporation.

Racking Circuit Breakers. Racking low-voltage power circuit breakers or medium-voltage vacuum circuit breakers in/out can be a hazardous work procedure. One safe work practice is to increase the working distance when performing this activity. If the working distance from the potential arc flash is increased to 36 inches rather than 18 inches, the potential incident energy at 36 inches is typically about one-fourth what it would be at 18 inches. The following methods can be used to move the worker farther from the potential arc source or outside the flash boundary for hazardous operations:

1. External and integrated remote-controlled motorized devices are available that rack in and out low- and medium-voltage circuit breakers. **See Figure 10-11.**
2. Extended length, hand-operated racking tools. **See Figure 10-12.**

Replacing Circuit Breakers. When adding a circuit breaker to a panel or replacing a circuit breaker, ensure that the circuit breaker is suitable for that panel and the circuit breaker has the proper ampere rating, voltage rating, interrupting rating,

Figure 10-12. RACKING TOOLS

Figure 10-12. *Racking tools of varying lengths are available to move the worker farther from the potential arc source.* Courtesy of DuPont.

and equivalent short-circuit performance. Circuit breaker frames that are part of the tested listing of the panel are identified on a label provided on the panel or switchboard. **See Figure 10-13.** Only circuit breaker frames identified on this label should be installed in the panel or switchboard.

With molded case circuit breakers, there typically is a variety of circuit breakers from the same manufacturer which are physically interchangeable but have different ratings. **See Figure 10-14.** These three circuit breakers are the same frame size and are physically interchangeable. The 240-V circuit breaker could physically be installed in place of the 600-V circuit breaker, the 100-A circuit breaker could physically be installed in place of the 20-A circuit breaker, and the 10-kA interrupting rated circuit breaker could physically be installed in place of the 65-kA interrupting circuit breaker. Switchgear, with draw-out power circuit breakers, have cell coding plates that prevent the installation of improperly rated circuit breakers.

Fuses

Several work practices should be evaluated when an electrical installation includes fuses. These include testing and replacing fuses.

Testing Fuses. When a fuse is suspected of having opened, safe work practices designated by the employer should be followed. One option is to deenergize

Figure 10-13. PANELBOARD LABELS

Devices To Be Installed Or Replacement Units Must Be From The Same Manufacturer Of The Same Type And Have Equal Or Greater Interrupting Ratings.

Any Space On This Panel Will Accept A Breaker Of The Same Frame Size As The Opposite Breaker.

Breaker - Switch Spaces On This Panel Will Accept One Of The Following Breakers or Switches Along With Their Respective Connector Kit.

SPACE	BREAKERS	CONNECTOR KIT CATALOG NO.
4-1/8 (104.8 mm)	1 POLE EHD/FD/FDB/HFD/FDC	KPRL4FD1
	2 POLE EHD/FD/FDB/HFD/FDC	KPRL4FD2
	3 POLE EHD/FD/FDB/HFD/FDC/FDE/HFDE/FDCE	KPRL4FD
	2 POLE ED/EDH/EDC/EDB/EDS	KPRL4ED2
	3 POLE ED/EDH/EDC/EDB/EDS	KPRL4ED
	3 POLE FD CURRENT LIMITER (Single)	KPRL4FCLIM
	2-3 POLE JD/JDB/HJD/JDC (Single)	KPRL4JDS
	2-3 POLE JD/JDB/HJD/JDC (Twin)	KPRL4JDT
	2-3 POLE FBP/FCL	KPRL4FBP
	2-3 POLE FDB - LFB	KPRL4FBP
5-1/2 (139.7 mm)	2-3 POLE DK/KD/KDB/HKD/KDC (Single)	KPRL4KDS
	2-3 POLE DK/KD/KDB/HKD/KDC (Twin)	KPRL4KDT *
	2-3 POLE LGE/LGS/LGH/LHH/LGC/LGU	KPRL4LG ***
8-1/4 (209.6 mm)	2-3 POLE LD/HLD/LDC	KPRL4LC
	2-3 POLE LA-P	KPRL4LAP
	2-3 POLE LCL	KPRL4LCL
	2-3 POLE MDS/HMDS/MDSY	KPRL4MC
	2-3 POLE MDL/HMDL	KPRL4MD
	2-3 POLE NB-P	KPRL4NBP *
	2-3 POLE ND/HND/NGS/NGH/NGC	KPRL4ND *
	2-3 POLE NHH	KPRL4ND

Figure 10-13. *Panelboards have labels which detail which type of circuit breakers are listed to be used in a specific panelboard.*

Figure 10-14. CIRCUIT BREAKER VARIETY

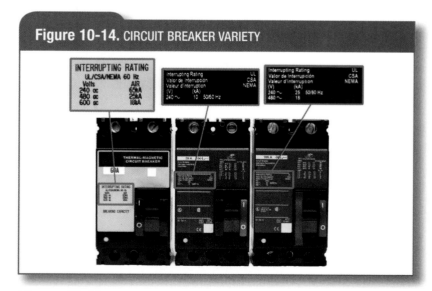

Figure 10-14. *These three circuit breakers are physically interchangeable, but have significantly different voltage ratings, interrupting ratings, and ampere ratings.*

Figure 10-15. TEST PROBES

Figure 10-15. *The test probes must be placed on the metal ferrules for a ferrule fuse.* Courtesy of Eaton's Bussmann Business.

continuity by resistance measurement. If a fuse is being replaced, the replacement fuse should be of a proper type and ampere rating.

When testing a fuse for resistance:
1. For ferrule fuses, place the test probes on the metal ferrules. **See Figure 10-15.**
2. For knife blade fuses, place the test probes on the metal blades, not on the fuse end caps. **See Figure 10-16.**

It is important to properly test knife-blade fuses. Fuse manufacturers do not generally design these types of fuses to ensure electrically energized fuse caps during normal fuse operation. For knife-blade fuses, electrical inclusion of the end caps into the circuit occurs as a result of the coincidental mechanical contact between the fuse cap and terminal extending through it. In most brands of knife-blade fuses, this mechanical contact is not guaranteed; therefore, electrical contact is not guaranteed. One fuse manufacturer has designed some knife-blade fuse versions so that the end caps are insulated to reduce the possibility of

the fuse from the source of power, including implementing lockout/tagout procedures, followed by removing both indicating and non-indicating fuses from the circuit and checking each fuse for

Figure 10-16. KNIFE BLADE FUSES

Figure 10-16. *For knife blade fuses, always place test probes on the metal blades as shown on the left. Do not place test probes on fuse end caps as shown on the right.* Courtesy of Eaton's Bussmann Business.

accidental contact with a live part. Thus a resistance reading to check for continuity (i.e., that the fuse is open or still usable) taken across the fuse caps is not indicative of whether the fuse is open. For knife-blade fuses, you should always have the test probes touch the metal knife blades. **See Figure 10-17.**

As part of diagnostic testing/troubleshooting, fuses can be checked via voltage testing while the equipment is energized. Safe work practices and proper personal protective equipment (PPE) must be utilized during this process. The test probes should contact the metal end caps of ferrule fuses and the blades on knife-blade fuses, much in the same way discussed earlier in this section.

Replacing Fuses. Fuses that interrupt a circuit should be replaced with the proper fuse in terms of both type and ampere rating. Modern current-limiting fuses are always recommended. When using modern current-limiting fuses, new factory-calibrated fuses are installed in the circuit, and the original level of overcurrent protection is maintained for the life of the circuit.

In most newer building systems and utilization equipment, the fuse mountings accept only current-limiting fuses of a specific Underwriters Laboratories Inc. (UL) Class fuse. For example, Class J fuse mountings will accept only UL Class J fuses; no other UL Class fuses can be installed in a Class J mounting. This standardization ensures a unique safety system in regard to voltage rating, interrupting rating, and short-circuit protection capabilities. The reason is that per the UL standard, all Class J fuses are rated as 600 V AC, have at least a 200,000-A interrupting rating, and provide a specific degree of current limitation (at a minimum) under short-circuit conditions. Thus, when a Class J fuse is

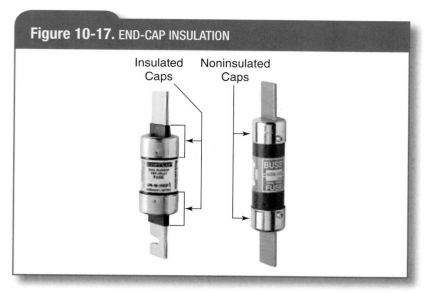

Figure 10-17. END-CAP INSULATION

Insulated Caps   Noninsulated Caps

Figure 10-17. The fuse on the left is designed to ensure that the end caps are electrically insulated. The fuse on the right does not have the designed-in end-cap insulation; however, an electrical connection between the fuse blades and end caps is not assured. *Courtesy of Eaton's Bussmann Business.*

replaced, this system ensures that only a Class J fuse can be inserted in its place—that is, it ensures that the fuse has a 600 VAC rating, at least a 200,000-kA interrupting rating, and a very high degree of current limitation. In essence, this standardization creates an electrical safety system for overcurrent protection that is unique only for current-limiting fuse classes. Each UL fuse class has its own unique dimensions and specific voltage rating, interrupting rating, and current-limiting performance per the UL standard for fuses.

For older systems, where the fuse clips can accept older-style fuses (Class H), it is recommended to store and use only modern current-limiting fuses (Class RK1) that also can be used in Class H fuse clips. See the section in the chapter covering design and upgrades for improved arc-flash mitigation by upgrading existing fuses.

Summary

Maintenance of electrical systems is required by *NFPA 70E* and directly affects worker safety. Improperly maintained equipment or lack of maintenance can create safety hazards for Electrical Workers. *NFPA 70E* requires the condition of maintenance to be considered in an arc flash risk assessment. Preventive maintenance of all electrical equipment is proven to increase reliability and safety for Electrical Workers. Lack of maintenance on OCPDs is of particular concern as it may affect the performance of an OCPD, thereby resulting in increased incident energy. What should be done if the condition of maintenance is inadequate? There is no single answer to this situation and the action taken may depend on the circumstances of a specific situation. For instance, it can be factored into a risk assessment procedure. Another alternative is change the work plan and establish an electrically safe work condition rather than work on the equipment while energized. The frequency of maintenance required for electrical equipment is best determined by reviewing the instruction book for the equipment. *NFPA 70B* also provides guidelines on recommended maintenance intervals and instructions on how to set up an electrical maintenance program.

Worker safety can be enhanced through the implementation of OCPD safe work practices. Various work practices can improve the level of safety when working on electrical equipment; these include not resetting circuit breakers or replacing fuses after a fault until it is safe to reenergize the system, properly evaluating circuit breakers prior to resetting them after a fault interruption, using remote racking devices for circuit breakers, replacing circuit breakers and fuses with devices of the proper type and rating, and testing fuses properly.

Review Questions

1. After a circuit is de-energized by the automatic operation of a circuit protective device, the circuit shall not be manually reenergized until it has been determined that the equipment and circuit can be __?__ energized.
   a. cautiously
   b. quickly
   c. safely
   d. slowly

2. After interrupting a fault approaching its interrupting rating, a __?__ must be inspected and tested in accordance with the manufacturer's instructions.
   a. circuit breaker
   b. fuse
   c. ground fault protection relay
   d. motor starter

3. The 2015 edition of *NFPA 70E* states that OCPDs shall be __?__ in accordance with the manufacturer's instructions or industry consensus standards.
   a. maintained
   b. replaced
   c. retrofitted
   d. sized

4. For the purposes of *NFPA 70E*, Chapter 2, maintenance shall be defined as preserving or restoring the condition of electrical equipment and installations, or parts of either, for the safety of __?__ who are exposed to electrical hazards.
   a. employees
   b. employers
   c. salespersons
   d. vendors

5. *NFPA 70E* 205.4 recommends but does not require that OCPD maintenance, tests, and inspections be documented.
   a. True
   b. False

6. When performing the hazard identification and risk assessment for a specific task and the maintenance condition of the OCPD is inadequate, one option may be to alter the work plan and work de-energized.
   a. True
   b. False

7. When replacing a molded case circuit breaker in a panelboard, besides making sure the circuit breaker is from the same manufacturer, an Electrical Worker should be sure that the replacement circuit breaker has the __?__.
   a. longest warranty
   b. lowest ampere rating for the load
   c. lowest price
   d. proper voltage rating, ampere rating, and interrupting rating

8. Why is the condition of maintenance important to the electrical worker in regards to safety?
   a. The owner has less unexpected equipment downtime.
   b. Improper maintenance of OCPDs can result in an actual arc flash incident having higher arc flash incident energy than determined by the arc flash risk assessment.
   c. Less call backs for service.

9. When racking a circuit breaker, one means for a worker to reduce the risk of arc flash is to __?__.
   a. lubricate the circuit breaker stabs as the circuit breaker disengages.
   b. stand on an insulating floor mat while using a hand racking tool.
   c. use a remote racking device.

10. What is not a recommended safe work practice for fuse testing?
   a. Place the test probes on the metal end caps of a knife blade fuse.
   b. Place the test probes on the metal ferrules of ferrule fuses.
   c. Place the test probes on the metal blades of knife blade fuses.

Electrical System Design and Upgrade Considerations

Chapter Outline

- Overcurrent Protective Device Considerations
- Additional Design for Safety Considerations

Chapter Objectives

1. Understand the important role that system design can play in eliminating or reducing electrical hazards.
2. Understand that choosing and installing overcurrent protective devices (OCPDs) is an important safety design consideration in regard to arc-flash and arc-blast hazards.
3. Identify other engineering and design practices that can enhance electrical safety for new systems and for upgrading of existing systems.

Chapter 11

Reference

1. *NFPA 70E®*, 2015 Edition

An electrician with approximately 10 years of experience was working at a power plant. At about 12:00 p.m., he was electrocuted while replacing a limit switch. The victim had been on vacation up to and including the day before the incident.

Around 11:30 a.m. that day, the victim received a briefing on the need to replace a limit switch. The victim supposedly took a normal lunch break from 11:45 a.m. to 12:15 p.m. At about 12:25 p.m., three workers saw the victim lying face up underneath a conveyor belt. On the ground next to the victim were a pack of cigarettes and two folded one-dollar bills.

The body had no signs associated with a fall and was not in a position that would result from a fall. Examination of the body showed electrical burns on the right hand, aspiration of food into the secondary and tertiary bronchi, a contusion on the left mastoid scalp, and an abrasion of the right mid-tibial area. The medical examiner concluded that the cause of death was electrocution.

The probable sequence of events follows: The victim was standing on the conveyor belt guard installing the new limit switch. Two of the three wires were connected, and the victim was connecting the last wire, the 220-volt energized wire. The wire coming out of the conduit was too short to reach the switch. The victim probably grabbed the wire with his right hand and attempted to pull it farther out of the conduit. As he did, the bottom of part of his hand contacted the limit switch. When the wire hit the upper part of his palm, a circuit was completed. Due to the relatively low voltage, the victim was not killed instantly; his heart probably went into arrhythmia. He probably felt uncomfortable, decided to get down and have a smoke, and died seconds later.

The major causal factor in this fatal incident was the failure of the victim to follow standard procedures for locking out electrical power. The failure apparently resulted from inattention by the victim rather than the difficulty of the procedure or a lack of time to do the job. An explanation for the failure could be a somewhat cavalier attitude of the victim toward relatively low voltages. The research team observed such an attitude in other electricians at the plant.

Source: For details of this case, see FACE Investigation #83PA08. Accessed July 10, 2012.

For additional information, visit qr.njatcdb.org Item #1201.

INTRODUCTION

Implementing safety-related design can significantly impact worker safety. If a system is designed so that an electrical hazard does not exist, then the worker will not have to contend with the hazard. Also, if the use of a specific design technique mitigates an electrical hazard to a much lower level, the workplace, although not hazard free, is still a safer environment. Designing for safety provides the opportunity to eliminate or minimize electrical hazards before a worker ever becomes exposed to them. Also, there are some design concepts that may not reduce the hazard level, but which may reduce the probability of an incident occurring.

Read *NFPA 70E* Informative Annex O Safety-Related Design Requirements which stresses the importance of facility owners having risk assessments performed during the design phase. While still in the design stage, various alternative design concepts can be considered and the most appropriate incorporated into the final design and subsequent installation.

There are safety related design concepts that can be implemented for existing systems resulting in eliminating or mitigating electrical hazards. It may be prudent to have older equipment retrofitted with modern devices or equipment thereby resulting in better electrical safety features which eliminate or mitigate hazards. Usually over the life-cycle of an electrical system, changes are implemented to expand or accommodate different needs. In some cases, these changes may result in more electrical hazards or increasing the level of hazards. An example is available fault currents in electrical systems that increase due to system changes. When a company does a facility-wide arc flash risk assessment at various locations within their electrical system by first conducting a facility-wide short-circuit current study, it is common to uncover many circuit breakers and fuses with inadequate interrupting rating. These situations are a serious safety hazard and corrective action should be taken immediately.

This chapter briefly mentions several techniques for new systems as well as upgrades for existing systems. However, there are many more design techniques beyond what is mentioned here and new ideas and products are continually being introduced. The reader is encouraged to continually strive to search for, investigate, and implement design techniques that result in safer workplaces. Even if a worker is not involved in the design process, they should mention viable design concepts and system upgrades to management, design personnel, and safety personnel to help those who are in the design process understand the value of various alternatives.

OVERCURRENT PROTECTIVE DEVICE CONSIDERATIONS

Choosing and installing overcurrent protective devices (OCPDs) is an important safety design consideration in regard to arc flash and arc blast. Overcurrent protective device selection decisions for new systems and for existing system upgrades may affect arc-flash hazards either positively or negatively. The magnitude of the arcing fault current and the length of time the arcing fault current flows are directly related to the arc flash energy released. OCPDs are the "safety valves" that may limit the magnitude of the arcing fault current that flows and reduce the time duration of the current, thereby potentially mitigating the arc-flash hazard. The three staged tests in the *Electrical Hazard Awareness* chapter illustrate this concept. Many considerations related to circuit breakers, fuses, and overcurrent relays can affect electrical safety. The selection of OCPDs can have a significant impact on the level of potential arc-flash incident energy.

The level of arc flash hazard for a piece of electrical equipment having a specific available fault current is directly related to the protecting OCPD's time-current characteristics. The chapter *Bolted and Arcing Fault Current and Reading Time-Current Curves* and the chapter *Methods to Accomplish the Arc Flash Risk Assessment* provide insight on this topic. There are a variety of time-current characteristics available for fuses, circuit breakers, and overcurrent relays. In the design stage, the selection of specific OCPD time-current characteristics along with the fault current will determine the arc flash risk assessments.

See Figure 11-1. This shows the time-current characteristic curves for two 600-A time-delay fuses. The green fuse curve represents a 600-A UL Class RK5 time-delay fuse. The blue curve represents a 600-A UL Class J time-delay fuse. For most applications needing a 600-A fuse, either one of these fuses could be used. Assuming an available short-circuit current of 16,400 A (black vertical arrow) with corresponding arcing fault current of 10,000 A (red vertical arrow), the 600-A Class RK5 fuse clears in approximately 0.085 seconds denoted by the red dot. The 600-A Class J fuse clears this same 10,000 A arcing fault in slightly less than 0.01 seconds (approximately ½ cycle or 0.0083 seconds) denoted by the yellow dot. This is just at the threshold of entering the current-limiting range of this Class J fuse. An arc flash risk assessment using an arcing fault current of 10,000A and the clearing times by these

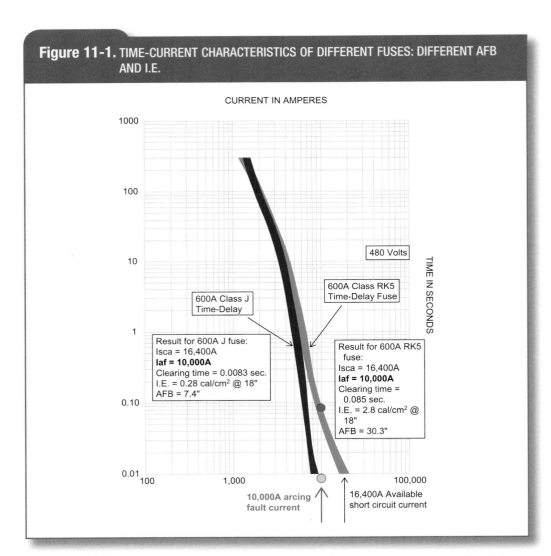

Figure 11-1. TIME-CURRENT CHARACTERISTICS OF DIFFERENT FUSES: DIFFERENT AFB AND I.E.

Figure 11-1. *Different time-current characteristics for these two fuses may result in different arc flash risk assessments for a specific arcing fault current.*

two fuses results in 2.8 cal/cm² for the 600-A Class RK5 fuse and 0.28 cal/cm² for the 600-A Class J fuse. Both scenarios are for 18" working distance. The arc flash boundaries are shown on the figures for each scenario.

Figure 11-2 represents the time-current curve for a 600-A electronic sensing molded case circuit breaker. **Figure 11-3** represents the time-current curve for 600-A low voltage power circuit breaker with a short-time-delay of 0.5 seconds. For an 11,850A arcing fault current the insulated case circuit breaker interrupts

in 0.04 seconds resulting in an incident energy calculation of 1.6 cal/cm² at 18" working distance and the LVPCB interrupts in 0.5 seconds resulting in an incident energy calculation of 19.95 cal/cm² at 18" working distance. These two figures illustrate the benefit of circuit breakers operating in their instantaneous trip range; the faster an OCPD clears an arcing fault the less the arc flash incident energy. The arc flash boundaries are shown on the figures for each scenario.

These examples illustrate that the selection of OCPDs can make a difference

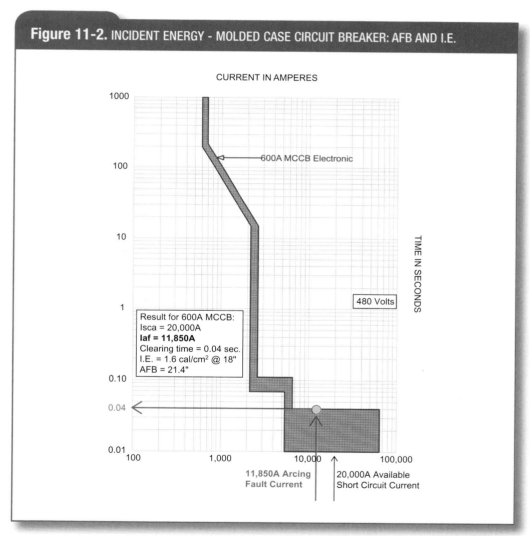

Figure 11-2. INCIDENT ENERGY - MOLDED CASE CIRCUIT BREAKER: AFB AND I.E.

Figure 11-2. This illustrates the resulting incident energy for a molded case circuit breaker with an instantaneous trip; compare this information to that illustrated in Figure 11-3.

Appendix E to § 1910.269, Protection From Flames and Electric Arcs, contains information to help employers estimate available heat energy as required by § 1910.269(l)(8)(ii), select protective clothing and other protective equipment with an arc rating suitable for the available heat energy as required by § 1910.269(l)(8)(v), and ensure that employees do not wear flammable clothing that could lead to burn injury as addressed by § 1910.269(l)(8)(iii) and (l)(8)(iv).

in the arc flash risk assessment. The incident energies in these examples were calculated using IEEE 1584 Edition 2002 *Guide for Performing Arc-Flash Hazard Calculations.*

Current-Limiting Overcurrent Protective Devices

A current-limiting overcurrent protective device interrupts a fault current in its current-limiting range in less than ½ cycle and reduces the fault current to something substantially less than would have flowed if the current-limiting device

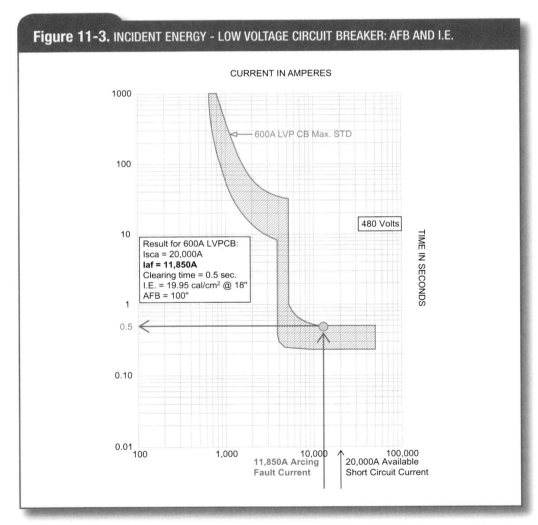

Figure 11-3. INCIDENT ENERGY - LOW VOLTAGE CIRCUIT BREAKER: AFB AND I.E.

Figure 11-3. This illustrates the resulting incident energy for a low voltage circuit breaker with a short-time setting and no instantaneous trip; compare this information to that illustrated in Figure 11-2.

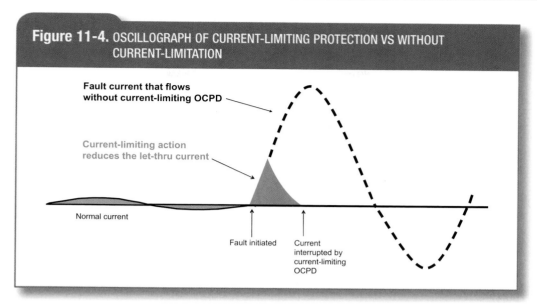

Figure 11-4. OSCILLOGRAPH OF CURRENT-LIMITING PROTECTION VS WITHOUT CURRENT-LIMITATION

Fault current that flows without current-limiting OCPD

Current-limiting action reduces the let-thru current

Normal current

Fault initiated

Current interrupted by current-limiting OCPD

Figure 11-4. The black dotted curve represents the fault current that flows with protection by a non-current-limiting OCPD and the red triangular shape represents the current flow by the current-limiting action of a current-limiting OCPD.

were not protecting the circuit. **See Figure 11-4.** The dotted curve represents the fault current that would flow without an OCPD. The red shaded triangular shape represents the current let-thru by a current-limiting OCPD. Overcurrent protective devices which are current-limiting offer superior arc flash mitigation when the available arcing fault current is in the current-limiting range of a current-limiting OCPD. When the fault current is in the current-limiting range of a current-limiting OCPD, the fault current's magnitude and time duration are both reduced, which reduces the energy released by an arcing fault. Referring to Chapter *Electrical Hazard Awareness* the section Staged Arc-Flash Tests (Tests 4, 3, and 1) provide illustrative examples of the value of current-limiting OCPDs.

Fuses and circuit breakers that are evaluated to be current-limiting per their respective UL product standard are marked on the product as "current-limiting." **See Figure 11-5.** As part of the UL product standard evaluation criteria, current-limiting fuses or current-limiting circuit breakers must perform to specific levels of maximum energy let-thru limits. Non-current-limiting fuses or circuit breakers do not have to perform to

Figure 11-5. CURRENT LIMITING FUSES AND CIRCUIT BREAKERS

VAC / 300DC OR
LPJ-60SP
Current Limiting

Bussmann
Low-Peak Fuses

LPJ-60SP

E·T·N
High Performance
Current Limiting
Circuit Breaker
E125C

Figure 11-5. 600 V or less overcurrent protective devices which are current-limiting are marked "Current Limiting." Courtesy of Eaton's Bussmann Business.

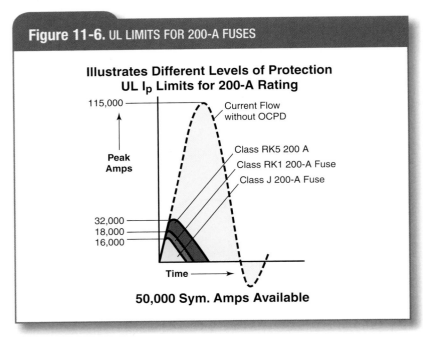

Figure 11-6. UL LIMITS FOR 200-A FUSES

Figure 11-6. UL I$_p$ limits for fuses rated as "200-ampere" illustrate the different degrees of current limitation and no current limitation.

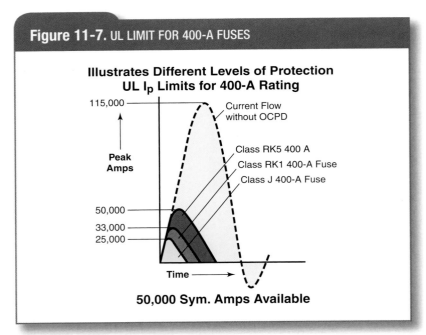

Figure 11-7. UL LIMIT FOR 400-A FUSES

Figure 11-7. UL I$_p$ limits for fuses rated as "400-ampere" illustrate the different degrees of current limitation and no current limitation.

specific maximum energy let-thru limits. **See Figure 11-6.** This illustrates the UL maximum energy let-through limits for four scenarios with an available fault current of 50,000A RMS symmetrical amperes: (1) current flow without OCPD (dotted black curve) (2) let-thru limit for a 200-A Class RK5 fuse (purple triangle), (3) let-thru limit for 200-A Class RK1 fuse (green triangle), and (4) let-thru limit for 200-A Class J fuse. Note that the dotted black line represents the asymmetrical short-circuit current that could flow with 50,000 symmetrical amperes available: the instantaneous peak current of that fault could reach 115,000 amperes. The UL fuse product standard instantaneous peak current limit for a 200-ampere Class RK5 fuse is 50,000 amperes; for a 400-ampere RK1 fuse, the limit is 33,000 amperes; and for a 400-ampere Class J fuse, it is 25,000 amperes. The lower the peak current and the smaller the triangle, the lower the energies released during and arcing fault event. **See Figure 11-7.** This depicts the UL maximum energy let-through limit values for 400-A overcurrent protective devices with an available fault current of 50,000A RMS symmetrical amperes. Notice that the various classes of 400-A fuses have greater let-thru currents than the 200-A fuses, which is to be expected.

It should be noted that the current-limiting performance criteria for circuit breakers is different than for fuses in their respective UL product standards; but this will not be covered in this publication. Those devices that are not listed and marked as current-limiting may not significantly reduce the level of fault current, and they may take longer to interrupt. As a consequence, non-current-limiting OCPDs may permit larger amounts of energy to be released during an arcing fault.

Different degrees of current limitation exist. Different OCPDs may become current limiting at different levels of short-circuit current. Moreover, once in the current-limiting range, different devices are more current limiting than others. If the arcing short-circuit current is in the current-limiting range of current-limiting fuses, the incident energy released during an arcing fault typically does not increase as the fault current increases. This is an important consideration in mitigating arc flash hazards. This means for certain fuse types and ampere ratings, high arcing fault currents are not a major issue since the fuse's current-limiting ability mitigates the arc flash energies so well.

Class RK5 fuses offer a good level of current-limiting protection. A better choice for applications using Class R fuse clips, however, are Class RK1 fuses; these fuses are more current limiting and enter their current-limiting range at lower fault levels. As a consequence, circuits protected by Class RK1 fuses will generally produce a lower arc flash incident energy than circuits protected by Class RK5 fuses.

Class J, Class CF, Class RK1, Class CC, and Class T fuses offer the best practical current-limiting protection. They generally have a significantly better degree of current limitation than other alternatives. They also typically enter their current-limiting range at lower currents than the other fuses or limiter alternatives. These types of fuses provide the greatest current limitation for general protection and motor circuit protection.

Overcurrent Protective Devices Life Cycle Consistency

Design criteria for choosing the type of OCPD for new installations or renovations to existing electrical systems should take into account whether the OCPDs will retain their specified fault-clearing operating characteristics over the life cycle of the system. This aspect depends on which type of OCPD is used and whether the appropriate level of maintenance resources, which depends on the type of OCPD, will be allocated throughout the electrical system's life cycle.

The reliability of OCPDs in retaining consistent fault-clearing performance over the system life cycle directly impacts arc flash hazards. *NFPA 70E* recognizes this relationship in 130.5(3) The arc flash risk assessment shall "take into consideration the design of the overcurrent protective device and its opening time, including its condition of maintenance" and in Informational Notes 1 and 4. IN No. 1 advises that inadequate maintenance of OCPDs can result in increased incident energy, while IN No. 4

points to Chapter 2 as the source of safety-related maintenance requirements. 205.4 requires maintaining OCPDs per the manufacturer's instructions or industry consensus standards and states that the inspections, maintenance, and tests must be documented. Other notable sections related to OCPDs include 210.5, 225.1, 225.2, 225.3, and 130.6(M).

The opening time of OCPDs is a critical factor for the resultant arc flash energy released when an arcing fault occurs. The longer an OCPD takes to clear a given arcing fault current, the greater the arc flash hazard. When an arcing fault or any short-circuit current occurs, the OCPD must be able to operate as intended. Therefore, the reliability of OCPDs is critical: they need to open as originally specified; otherwise, the arc flash hazard can escalate to higher levels than expected.

Two different types of overcurrent protection technology provide different choices in maintenance requirements and might affect the arc flash hazard.

OSHA Tip

Appendix E to § 1910.269 - Protection From Flames and Electric Arcs

III. Protection Against Burn Injury

A. Estimating Available Heat Energy

Calculation methods. Paragraph (l)(8)(ii) of § 1910.269 provides that, for each employee exposed to an electric-arc hazard, the employer must make a reasonable estimate of the heat energy to which the employee would be exposed if an arc occurs. Table 2 lists various methods of calculating values of available heat energy from an electric circuit. The Occupational Safety and Health Administration does not endorse any of these specific methods. Each method requires the input of various parameters, such as fault current, the expected length of the electric arc, the distance from the arc to the employee, and the clearing time for the fault (that is, the time the circuit protective devices take to open the circuit and clear the fault). The employer can precisely determine some of these parameters, such as the fault current and the clearing time, for a given system. The employer will need to estimate other parameters, such as the length of the arc and the distance between the arc and the employee, because such parameters vary widely.

Current-Limiting Fuses. Current-limiting fuses are reliable and retain their ability to open as originally designed under fault conditions. **See Figure 11-8.** When a fuse is replaced, a new factory-calibrated fuse is put into service, and the circuit has reliable protection with performance equal to the original specification. Modern current-limiting fuses do not require maintenance other than visual examination and ensuring that there is no damage to fuses from external thermal conditions (such as conductor terminations), liquids, or physical abuse. It is important that when fuses are replaced, the proper type and ampere rating are used.

Circuit Breakers. Circuit breakers are mechanical OCPDs that require periodic exercise, maintenance, testing, and possible replacement. A circuit breaker's reliability and operating speed depend on its original specification and its condition. A specific circuit breaker's condition of maintenance is influenced by variables, including the following:

- Length of service
- Number of manual operations under load
- Number of operations due to overloads
- Number of fault interruptions
- Humidity
- Condensation
- Corrosive substances in the air
- Vibrations
- Invasion by foreign materials or liquids
- Thermal damage caused by loose connections
- Erosion of contacts
- Erosion of arc chutes
- Maintenance history

A circuit breaker manufactureur's instructions for maintenance should be followed. In addition, 225.3 requires a circuit breaker that interrupts a fault near its interrupting rating to be inspected and tested.

Consideration for Fusible Systems

In addition to current limitation and consistency considerations, there are other

Figure 11-8. CURRENT-LIMITING FUSES

Figure 11-8. *A range of modern current-limiting fuses are shown: 30 A, 600 V Class CC; 30 A, 600 V Class J; 30 A, 600 V Class RK1; 100 A, 250 V Class RK1; 100 A, 600 V Class RK1; and 1200 A, 600 V Class L. Courtesy of Eaton's Bussmann Business.*

important considerations for fusible systems.

Rejection-Style Class J, CF, T, R, G, and L Fuses. It is important to ensure that arc flash protection levels are maintained as a facility ages. Class J, CC, CF, T, R, and L fuses provide an advantage in that these fuse classes each having physical size/mounting provisions that are unique to each fuse class. **See Figure 11-9.** A fuse of a given class cannot be inserted into mountings designed for another class. As a consequence, fuses with lower voltage ratings, interrupting ratings less than 200,000 A (100,000 A for Class G fuses), or lower current limitation performance cannot be accidentally put into service.

However, the arc-flash incident energy can vary for a specific ampere rated fuse of a given class. For a given ampere rated fuse of a specific class there may be more than one type fuse; such as, Class J fuses are available with non-time delay characteristics and also time-delay characteristics. Each can have different arc-flash mitigation results. In addition, there may be differences in performance by fuses of different manufacturers. If an arc-flash incident energy calculation has been performed for a circuit based on a specific type and manufacturer's fuse, it is recommended that replacements should be a fuse of the same type and manufacturer.

The exception: if the replacement fuse's time-current characteristics are equal or faster clearing time than the original fuse, then the substitution should provide similar or better arc flash mitigation results.

Enhanced Safety of New-Style Fusible Disconnects. In the past decade, the new fusible disconnects have become even

Figure 11-9. REJECTION-STYLE FUSES

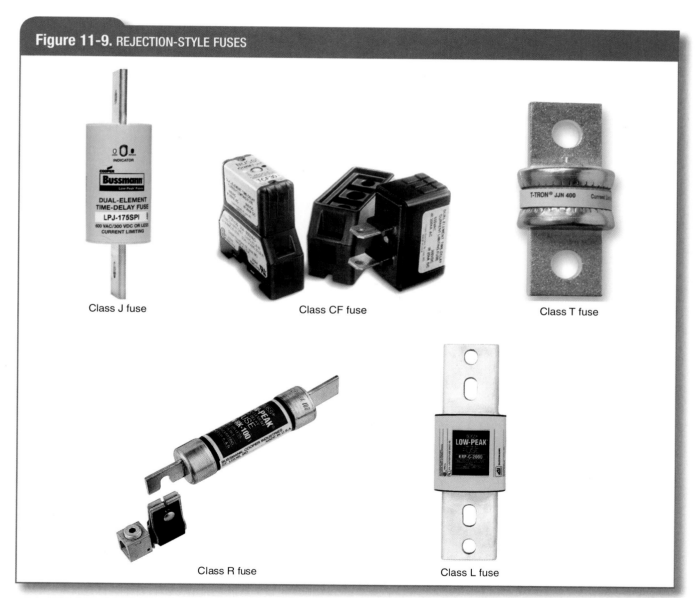

Class J fuse

Class CF fuse

Class T fuse

Class R fuse

Class L fuse

Figure 11-9. Examples of rejection-style fuses are shown. Courtesy of Eaton's Bussmann Business.

safer. **See Figure 11-10.** For instance, the fusible disconnects in this panelboard permit servicing of the fuses without opening the panel trim. The disconnect is interlocked with the fuses. When extracting or inserting a fuse, place the disconnect handle in the "OFF" position.

Switches Equipped with ARMS. Arc Reduction Maintenance Switching (ARMS) is one of the most important new technology options to include on larger fusible switches (1200A and greater) to mitigate the arc flash hazard for lower available arcing fault currents situations. **See Figure 11-11.** When the ARMS option is switched "on", if an arcing fault event were to occur on a circuit with this fus-

ible switch protecting, there is additional enhanced arc flash protection for a lower range of arcing fault currents. With the ARMS added to a fusible switch the result of the ARMS and fuses together can be significantly lower incident energy levels for a wide range of arcing fault currents from low magnitude to high magnitude. There is an option where the ARMS can be locked into the "on" position which provides the provisions for lockout/tagout.

The common use of an ARMS is when a worker will be interfacing with electrical equipment that is not in an electrically safe work condition, the ARMS is switched "on" prior to the worker approaching the equipment. By this action, the potential arc flash risk assessment may be less while the worker checks for the absence of voltage or works on energized equipment. When the work is completed, the ARMS is set back to the "off" mode to provide overcurrent protection via only the normal fuse time-current curve characteristic.

Figure 11-10. NEW STYLE FUSIBLE PANELBOARD

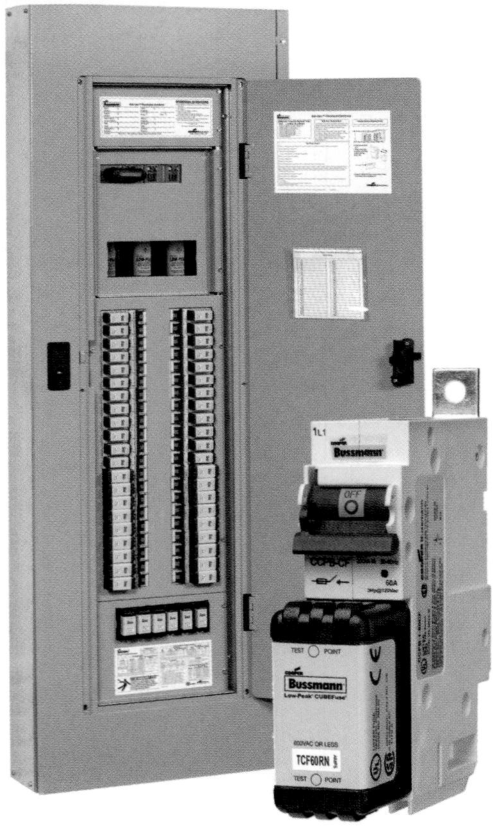

Figure 11-10. *In this fusible panelboard, the branch circuit fuses can be serviced without opening the panel trim. Courtesy of Eaton's Bussmann Business.*

Figure 11-11. FUSIBLE SWITCH EQUIPPED WITH ARMS

Figure 11-11. *A fusible switch equipped with ARMS will mitigate arc flash hazard. Courtesy of General Electric Corp.*

Figure 11-12A shows a 480-V scenario where a 1200-A fused switch (without ARMS protection) is protecting downstream equipment which has an available short-circuit current of 20,000 A. An arc fault risk assessment using the incident energy calculation method would result in an arcing fault current of 11,850 A and fuse clearing time of 0.30 seconds. The incident energy calculation equals 12 cal/cm^2 at 18 inches working distance and AFB calculation equals 73 inches.

Figure 11-12B is the same situation as used for Figure 11-12A, except this

assessment includes the added protection of with the ARMS being switched "on". The available arcing fault current is a magnitude where the addition of an ARMS is a benefit. In this case, the ARMS clears the 11,850-A arcing fault current in 0.05 seconds resulting in an incident energy calculation of 2 cal/cm^2 at 18 inch working distance and AFB calculation equals 24.5 inches. The incident energies in these examples were calculated using IEEE 1584 Edition 2002 *Guide for Performing Arc-Flash Hazard Calculations*.

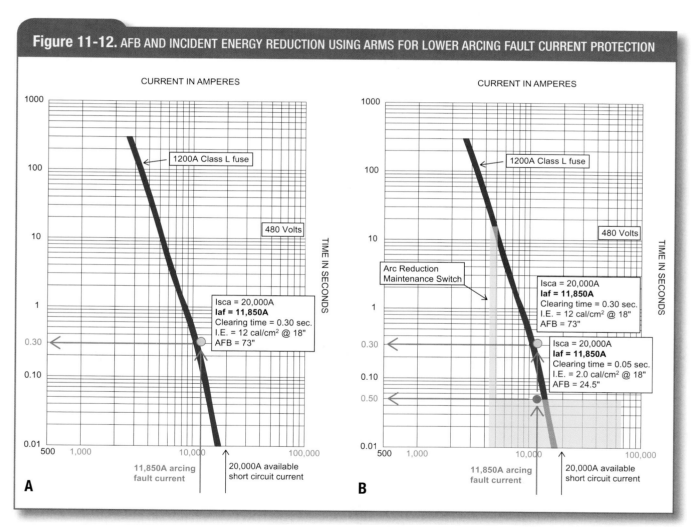

Figure 11-12. AFB AND INCIDENT ENERGY REDUCTION USING ARMS FOR LOWER ARCING FAULT CURRENT PROTECTION

***Figure 11-12**. A: This illustrates the incident energy and AFB calculation results with 1200-A fuses with 20,000 A available short-circuit current; key information is the arcing fault current and clearing time of the fuse for that arcing fault current. B: This illustrates the reduction compared to results in Figure 11-12A for the incident energy and AFB calculations by using an arc reduction maintenance system. Courtesy of Eaton's Bussmann Business.*

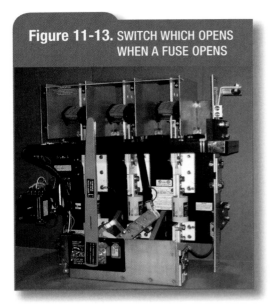

Figure 11-13. SWITCH WHICH OPENS WHEN A FUSE OPENS

Figure 11-13. Switches with the capability to automatically open when a fuse opens can help reduce the arc flash incident energy for larger ampere-rated circuits. Courtesy of Boltswitch.

Figure 11-14. FUSES TO REDUCE ARC FLASH INCIDENT ENERGY

Renewables (Class H)

Dual Element Time Delay (Class RK1)

Fast Acting (Class RK1)

One-Time (Class H)

Dual Element Time Delay (Class RK5)

Figure 11-14. Replacing all existing Class H, Class K, and Class RK5 fuses with Class RK1 fuses can mitigate the arc-flash hazard to lower incident energies in many situations.

Switches for Circuits 800 Amperes and Higher. Switches with a shunt-trip that opens all three poles of the switch when the first fuse opens reduce arc flash energy levels. **See Figure 11-13.** This option can be included on new switches or can be retrofitted on some existing switches. Tests have shown that it can reduce the arc flash risk assessment on circuits with large ampere ratings. If a switch is used in this manner, it should undergo periodic maintenance to ensure consistent arc flash performance. This approach does not mitigate the arc flash incident energy to as low of level as the fusible switches equipped with ARMS for the lower magnitude arcing faults.

Reduce Arc Flash Risk Assessment for Existing Fusible Systems. If the electrical system is an existing fusible system, consider replacing or upgrading the existing fuses with fuses that are more current limiting. This step can reduce the arc flash risk assessment and is a common practice.

Owners of existing fusible systems should consider upgrading Class H, K5, K9, and RK5 fuses to Class RK1 fuses, and should verify that Class J and Class L fuses are the most current-limiting models available. An assessment of many facilities will reveal that the installed fuse types are not the most current limiting, or that fuses were installed decades ago and new fuses with better current limitation are now available. **See Figure 11-14.**

Circuit Breakers

In addition to consistency considerations that have already been discussed for circuit breaker systems, other important points must be considered in relation to electrical safety.

Arc Flash Reduction Maintenance System. This is one of the most important options to include on larger circuit breakers to mitigate the arc flash hazard. **See Figure 11-15.** Arc flash reduction maintenance system (ARMS) for circuit breakers is a recent technology that improves electrical safety for workers who must work

Figure 11-15. CIRCUIT BREAKER EQUIPPED WITH ARMS

Figure 11-15. A circuit breaker equipped with an arc flash-reducing maintenance switch is controlled by a screw driver. Courtesy of Eaton Corporation.

on equipment that is not in an electrically safe work condition. When the ARMS is switched "on", if an arcing fault event occurs on a circuit which the circuit breaker protects, the circuit breaker trips without intentional delay. Even if a circuit breaker normally has a short-time-delay setting, the ARMS when "on" overrides the short time-delay and trips as fast as it can. In addition, the fault current pickup setting of the ARMS may be set to a lower current pickup current level than the normal instantaneous trip or short-time delay setting. The result can be significantly lower incident energy, if an arcing fault occurs.

The common use of an ARMS is when a worker will be interfacing with electrical equipment that is not in an electrically safe work condition, the ARMS is switched "on" prior to the worker approaching the equipment. By this action, the potential arc flash hazard is less while the worker checks for the absence of voltage or works on energized equipment. When the work is completed, the ARMS is set back to the "off" mode to provide overcurrent protection via the normal circuit breaker settings.

Some ARMS systems include a manual switch that the worker uses to turn the bypass on and off. Other ARMS systems use automatic means to sense whether the worker is within close proximity of the equipment and then automatically turn the arc reduction maintenance system on and off. For instance, as a worker opens an enclosure, a motion sensor may detect the person and invoke an ARM setting. There typically are options which permit the ARMS to be locked in the "on" position. ARMS can be retrofit to existing circuit breakers in many cases.

The 2014 *NEC* 240.87 requires technology that reduces the potential arc flash energy while personnel work on equipment protected by a circuit breaker which can be adjusted to a continuous current rating of 1200 A or higher. A circuit breaker equipped with an ARMS provides a reliable means for compliance.

Background

Arc Reduction Maintenance System Performance Varies

ARMS technology is implemented in various ways by circuit breaker manufacturers and it is advisable when specifying or buying circuit breakers with ARMS to investigate this aspect. It can make a significant difference in the arc flash incident energy and AFB calculations. For instance, for electronic sensing circuit breakers, in some cases the ARMS function utilizes a separate analogy magnetic system which may be much faster in clearing time for an arcing fault than if the ARMS is implemented via electronic sensor. Slower reaction time results in longer clearing time for an arcing fault with an end result of a greater arc flash hazard.

Figure 11-16. AFB AND I.E. FOR CIRCUIT BREAKER WITHOUT ARMS

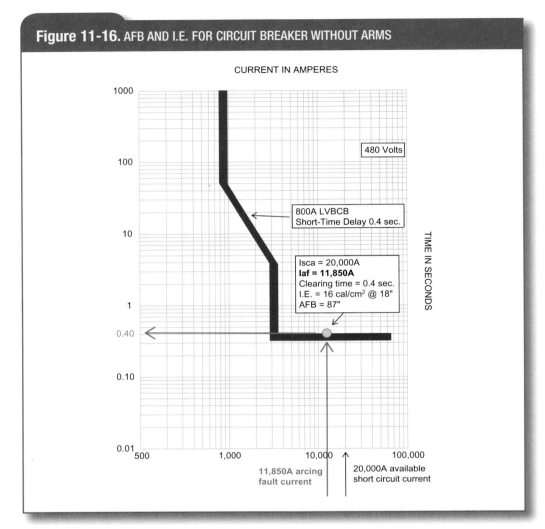

CURRENT IN AMPERES

480 Volts

800A LVBCB
Short-Time Delay 0.4 sec.

Isca = 20,000A
Iaf = 11,850A
Clearing time = 0.4 sec.
I.E. = 16 cal/cm² @ 18"
AFB = 87"

TIME IN SECONDS

11,850A arcing
fault current

20,000A available
short circuit current

Figure 11-16. This illustrates the incident energy and AFB calculation results with a 1200-A circuit breaker having a short-time delay and 20,000 A available short-circuit current; key information are the arcing fault current and clearing time of the circuit breaker for that arcing fault current. Compare results to those in Figure 11-17.

See Figure 11-16. This is the time-current curve for an 800-A low voltage power circuit breaker. It has a short-time delay setting of 0.4 seconds and it has no instantaneous trip. In this example with an available short-circuit current of 20,000 A, the arcing fault current would be calculated at 11,850 A. The circuit breaker opening time for 11,850 A arcing fault current is 0.4 seconds resulting in calculations of a 16 cal/cm² arc flash incident energy at 18 inches working distance and 87 inches arc flash boundary.

By incorporating an ARMS with this circuit breaker the arc flash incident energy and AFB could be considerably lowered. **See Figure 11-17.** This shows the same 800A circuit breaker equipped with an ARMS set at 2X pickup. If the ARMS is switched to the "on" position and this same arcing fault current of 11,850A occurred, the arc flash incident energy would be calculated as 1.6 cal/cm² at 18 inches and the arc flash boundary would be 21.4 inches. This arc flash risk assessment is so much better due to the

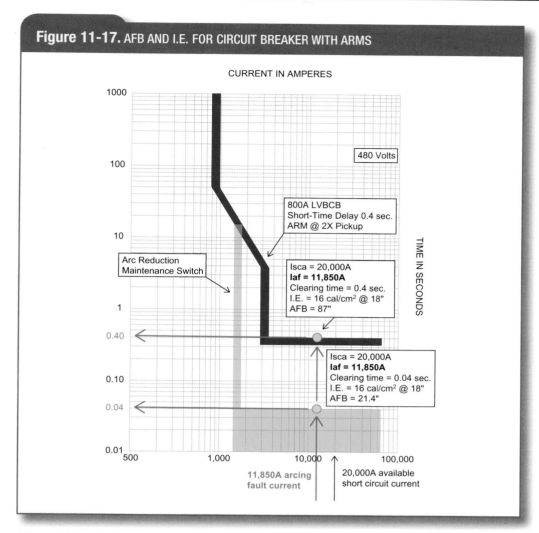

Figure 11-17. AFB AND I.E. FOR CIRCUIT BREAKER WITH ARMS

Figure 11-17. This illustrates the reduction compared to results in Figure 11-16 for the incident energy and AFB calculations by using an arc reduction maintenance system.

ARMS clearing the arcing fault current in 0.04 seconds.

Adjusting Circuit Breaker Instantaneous Trip Settings. Some circuit breakers have adjustable instantaneous trip settings. **See Figure 11-18.** Such a setting allows adjustment of the fault level at which the circuit breaker will start to operate in its instantaneous mode. This adjustment traditionally is used to help avoid nuisance opening on normal load surges or to improve coordination. If the arcing fault current calculated falls in the current range between the instantaneous trip low and high pickup settings, adjusting the instantaneous trip setting can greatly impact the arc flash incident energy. If the instantaneous trip level is set too high, the circuit breaker might not operate in its instantaneous mode during an arcing fault. The result is a longer opening time, which in turn means that a much higher level of energy may be released during an arcing fault. Conversely, if the instantaneous trip is set low enough to operate

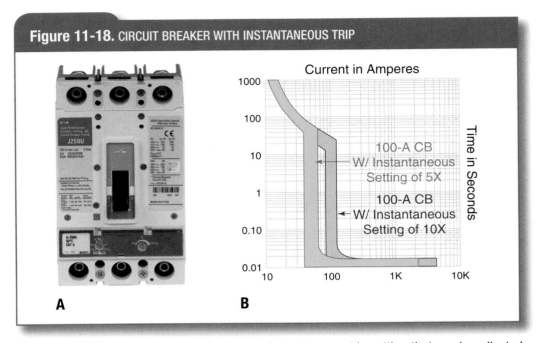

Figure 11-18. CIRCUIT BREAKER WITH INSTANTANEOUS TRIP

A

B

Figure 11-18. A: This circuit breaker has an instantaneous trip setting that can be adjusted by dials. B: A 100-A circuit breaker time-current curve with the instantaneous trip dial setting at 5X = 500 amperes and a different dial setting at 10X = 1000 amperes is illustrated. Left: Courtesy of Eaton Corporation.

for a low arcing fault condition, the arc flash incident energy may be mitigated to a lower level. An assessment is necessary to determinethe instantaneous trip settings adjusted to be as low as possible without incurring nuisance tripping.

One consideration when deciding whether to lower a circuit breaker's instantaneous trip setting is the effect this change will have on selective coordination of the OCPDs. Selective coordination is required in the *NEC* (620.62, 700.28, and other sections) for some circuits supplying life-safety loads. Where not required by the *NEC*, an owner may still want selective coordination in order to have more reliable power supply and avoid unnecessary power outages.

Sometimes maintenance or service personnel will adjust the instantaneous trip settings without approval. For this reason, it is recommended to perform arc flash risk assessments based on the highest possible instantaneous trip setting for a given circuit breaker.

Short-Time Delay Circuit Breakers. Due to *NEC* 240.87, 1200A circuit breakers and larger must have a means to mitigate the arc flash incident energy with an arc reduction maintenance system being the most easily implemented and cost effective technology. Some existing installed low voltage power circuit breakers (LVPCB), as well as some newly installed LVPCBs less than 1200A are equipped with a short-time delay setting (no instantaneous trip). This feature is intended to delay operation of the circuit breaker under fault conditions. **See Figure 11-19.** Circuit breakers with short-time delay (and no instantaneous trip) are used on feeders and mains so that downstream molded case breakers can clear a fault without tripping the larger upstream circuit breaker. Under fault conditions, a short-time delay sensor intentionally delays signaling the circuit breaker to open for the time duration setting of the short-time delay. Therefore, a fault is permitted to flow for an extended time. A low-voltage power circuit breaker with a short-time delay, and without instantaneous trip, permits a fault to flow for the length of time of the short-time delay setting, which might be 6, 12, 18, 24, or 30 cycles.

If an arcing fault occurs on the circuit protected by a device with a short-time delay setting, significantly more incident energy might be released while the system waits for the circuit breaker short-time delay to time out. The longer an OCPD takes to open, the greater the flash hazard due to arcing faults. Experience shows that the arc flash incident energy increase is directly proportional with the amount of time arcing current is permitted to flow.

System designers and users should understand that using circuit breakers with short-time delay settings (without instantaneous trip) could greatly increase the arc flash incident energy. If an incident occurs when a worker is at or near the arc flash source, that worker might be subjected to considerably more arc flash energy than if an instantaneous trip circuit breaker ora circuit breaker equipped with an ARMS is utilized. For existing circuit breakers with short-time delay settings and no instantaneous trip, arc reduction maintenance system retro-fitted to the circuit breaker is a viable solution to mitigate the incident energy. For new circuit breakers with short-time-delay and no instantaneous, purchase one that has ARMS as part of the circuit breaker.

Zone-Selective Interlocking. While utilizing short-time delay settings may be necessary to achieve selective coordination for a circuit breaker design, short-time delay settings can permit high incident energy levels under arcing fault condi-

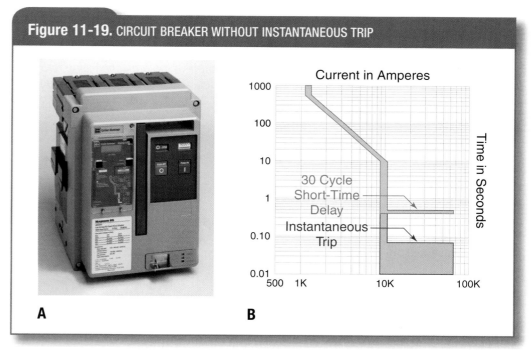

Figure 11-19. CIRCUIT BREAKER WITHOUT INSTANTANEOUS TRIP

A   **B**

Figure 11-19. A: A low-voltage power circuit breaker (LVPCB) is shown. B: The pink curve represents a LVPCB with a 30-cycle short-time delay setting and no instantaneous trip. The purple curve represents an instantaneous trip for a circuit breaker. A circuit breaker with an instantaneous trip will clear an arcing fault in less time than a circuit breaker with a short-time delay. Left: Courtesy of Eaton Corporation.

Figure 11-20. ZSI: FAULT IN CIRCUIT BREAKER 3 PROTECTION ZONE

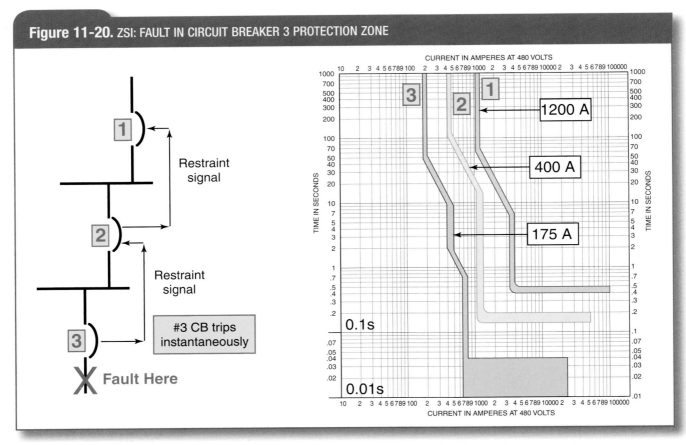

Figure 11-20. *The fault occurs on the loadside of circuit breaker 3 (175 A). This fault is in the protection zone for circuit breaker 3, and it operates instantaneously as the time-current curve above illustrates. When circuit breaker 3 senses a fault, it sends a restraint signal to circuit breakers 2 (400 A) and 1 (1,200 A); circuit breakers 2 and 1 then function with the short-time delay settings shown by the adjacent time-current curves.*

tions to occur. Another possible solution to this dilemma is to use circuit breakers equipped with zone-selective interlocking. Zone-selective interlocking is an option for some circuit breaker systems where circuit breakers communicate with each other and the time-current functions of the circuit breakers are dynamically modified based on the location of the fault condition. Circuit breakers with this option are equipped with communication wiring between the circuit breakers. The benefit is that when a downstream fault is not in a circuit breaker's protection zone, the circuit breaker can be set to operate with a short-time delay to provide selective coordination with downstream circuit

breakers. When a fault is within a circuit breaker's protection zone, then that circuit breaker overrides its short-time delay and operates as fast as it can, thereby providing better equipment protection and less incident energy under arcing fault conditions. Note: operating as fast as it can does not necessarily mean the circuit breaker will operate as fast as an instantaneous trip. In order to achieve selective coordination, the circuit breakers must be set to selectively coordinate before factoring in zone selective interlocking. **See Figures 11-20, 11-21, and 11-22.** The three fault scenarios for a system demonstrate three circuit breakers equipped with zone-selective interlocking (ZSI) and indicate

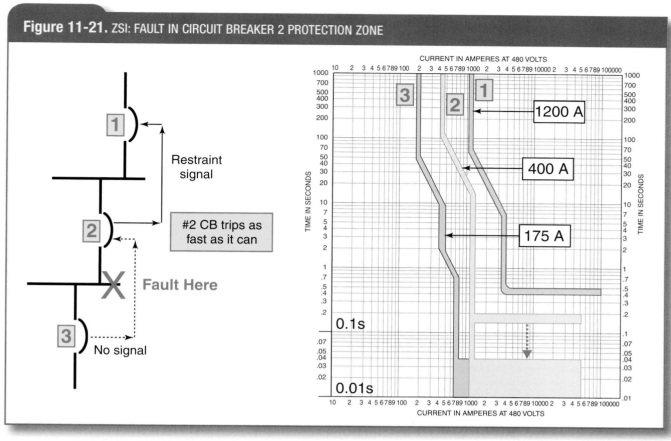

Figure 11-21. ZSI: FAULT IN CIRCUIT BREAKER 2 PROTECTION ZONE

Figure 11-21. For faults between circuit breakers 2 and 3 (circuit breaker 2 protection zone), circuit breaker 3 does not sense the fault current, so it does not send a restraint signal to circuit breaker 2. Circuit breaker 2 does sense the fault current and sends a restraint signal to circuit breaker 1, so that circuit breaker 1 functions with the short-time delay setting shown by the time-current curve. Circuit breaker 2 will function without any intentional delay; it opens as quickly as possible because it did not receive a restraint signal from circuit breaker 3.

how they can be selectively coordinated using short-time delays and still mitigate the arc-flash incident energy to lower levels. Each scenario shows the fault at a particular point in the system and the resulting effective time-current curve for that scenario. Note these three circuit breakers have to selectively coordinate prior to utilizing the ZSI function.

Relaying Schemes

Relay schemes can be used on both new installations and existing systems to mitigate arc flash hazard. When an arc flash event occurs that causes the relay to call for the circuit to be interrupted, the relay output signals a properly rated discon-

necting means to interrupt the circuit. The disconnecting means can be a vacuum interrupter, circuit breaker, or disconnect switch with a shunt trip.

Overcurrent Relays. Overcurrent relays are used for some low-voltage applications, but are more commonly used on medium- and high-voltage systems. The current and time settings to operate are important factors influencing the arc flash hazard with such systems. Coordination for medium- and high-voltage OCPDs is generally more critical from a business continuity basis, because an outage at this level represents a large amount of power. Thus the objective is

Figure 11-22. ZSI: FAULT IN CIRCUIT BREAKER 1 PROTECTION ZONE

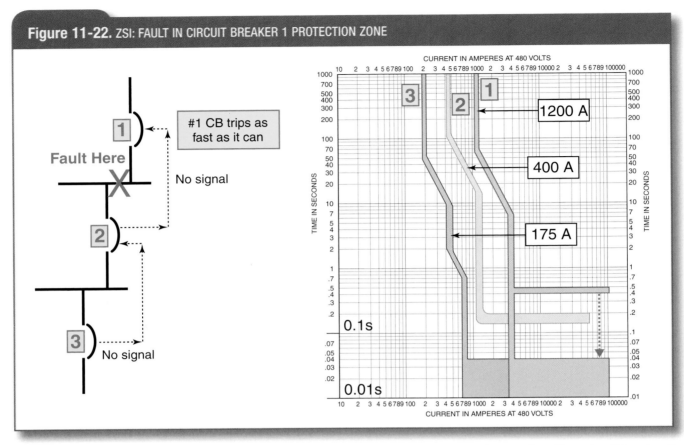

Figure 11-22. For faults between circuit breakers 1 and 2 (circuit breaker 1 protection zone), neither circuit breaker 3 nor circuit breaker 2 senses the fault, and no restraint signal is sent to circuit breaker 1. Because circuit breaker 1 senses a fault and does not receive a restraint signal, it will function without any intentional delay. The time-current curve shows circuit breaker 1 opening as quickly as possible, reducing the arc-flash hazard.

to set the relays to minimize the arc flash risk assessment as well as to provide acceptable coordination levels.

***Arc-Flash Relays.* See Figure 11-23.** Arc-flash relays can be utilized in conjunction with an interrupting disconnect device to mitigate incident energy levels by limiting the time of the fault. Typically these relays monitor for uncharacteristic current, and several light sensors placed at key locations inside an electrical enclosure. Other parameters that could be monitored include sound and pressure. If an arc-flash event occurs within the enclosure, the light, sound, and pressure all rapidly escalate to high levels. When

an uncharacteristic current flow is combined with one or more of these other parameters, the relay reaches its trip point and then signals a disconnecting means to interrupt. This method is an effective arc flash mitigation means for equipment protected by OCPDs that are too large in ampacity to quickly respond to the arcing fault current.

On a 480-volt service or unit substation with a large ampacity main OCPD, the arc-flash relay technology can be used to signal a medium-voltage disconnect and allow for fast interruption of an arc-flash event. However, this technology using light sensors typically may not be suitable with circuit breakers since dur-

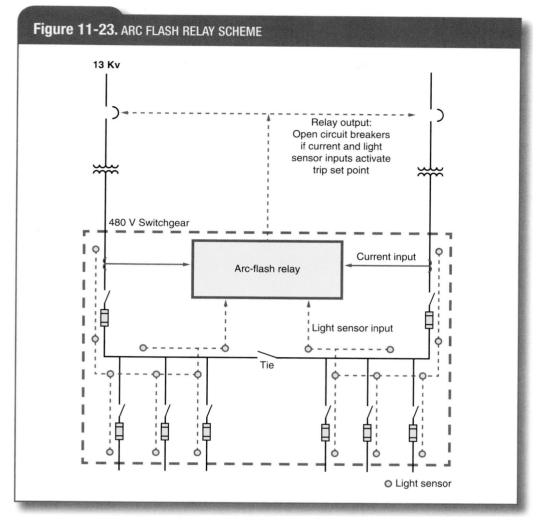

Figure 11-23. ARC FLASH RELAY SCHEME

13 Kv

Relay output:
Open circuit breakers
if current and light
sensor inputs activate
trip set point

480 V Switchgear

Arc-flash relay

Current input

Light sensor input

Tie

O Light sensor

Figure 11-23. *An arc flash relay scheme for double-ended switchgear can mitigate the incident energy to a lower level than is possible with standard OCPDs.*

ing an overcurrent interruption process circuit breakers may emit significant light via the venting process.

Separate Enclosure for Main Disconnect

Service entrance equipment typically has the most hazardous arc flash risk assessments due to the line-side connection being protected upstream by a slow operation utility OCPD. To reduce the arc-flash hazard exposure, the design may call for the main disconnect to be placed in a separate enclosure from the feeder discon-

nects or overcurrent protective devices. **See Figure 11-24.** With this configuration, it is possible to open the main disconnect and put the feeder enclosure in an electrically safe work condition. This requires the appropriate personal protective equipment (PPE) for this procedure; however, the arc flash rating of the PPE will be based on the main OCPD, not the utility OCPD. After an electrically safe work condition is achieved, workers do not need PPE and can work in the feeder enclosure without any risk of shock or arc flash. If the main is located in the same

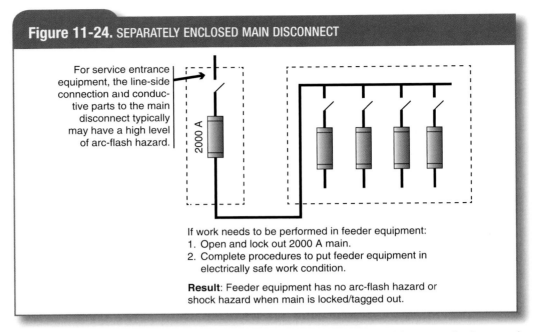

Figure 11-24. SEPARATELY ENCLOSED MAIN DISCONNECT

For service entrance equipment, the line-side connection and conductive parts to the main disconnect typically may have a high level of arc-flash hazard.

2000 A

If work needs to be performed in feeder equipment:
1. Open and lock out 2000 A main.
2. Complete procedures to put feeder equipment in electrically safe work condition.

Result: Feeder equipment has no arc-flash hazard or shock hazard when main is locked/tagged out.

Figure 11-24. A separate enclosure for the main disconnect can provide means for improved work practices for work in the downstream enclosure.

line-up, then even with the main disconnect open, there is typically an arc-flash hazard from the energized main disconnect, line-side connections, and conductive parts that are within the enclosure. If work is necessary within the main disconnect enclosure, it will be necessary to deenergize and lock out/tag out any disconnect upstream, which involves having the utility deenergize the service. A design featuring separately enclosed disconnect supply panels can also be used for industrial control panels (a separately enclosed disconnect is attached on the outside of the industrial control panel enclosure), motor control centers, branch panels, and distribution panels. This arrangement provides an electrical system designed for safer work practices.

If the arc-flash hazard in the feeder enclosure is higher than desired due to the ampacity and characteristics of the main OCPD, an arc flash relay (like that described earlier) can be used to reduce the arc-flash hazard. The arc-flash relay can have optic sensors in the feeder enclosure and use the main disconnect (which needs a shunt-trip) as the device that interrupts the current upon the relay sensing an arc flash.

Specify a Main on Each Service

Generally, a single main service disconnect provides for safer work practices than the six-disconnect rule for service entrances permitted in *NEC* 230.71. The six-disconnect rule is intended to reduce the cost of the service equipment, but this choice typically increases worker exposure to hazards. Without a main OCPD, the main bus and line terminals of the feeders are unprotected. In addition, equipment installed under the six-disconnect rule does not allow the bus to be deenergized without the utility being called to deenergize its supply.

If a worker must work in the enclosure of one of the feeders, the compartment can be placed in an electrically safe work condition. To achieve that end, typically a main disconnect should be locked out. Afterward, no energized conductors will remain in the feeder device compartment. (There may still be an arc-flash hazard due to the main line-side connection unless the main is in a separate enclosure.) **See Figure 11-25.** Note that the system represented by the single-line diagram on the left has a lower arc-flash hazard level than the system represented by the one-line diagram

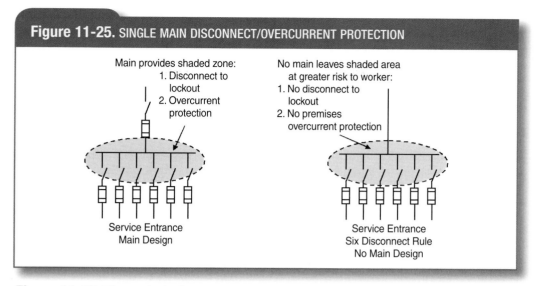

Figure 11-25. SINGLE MAIN DISCONNECT/OVERCURRENT PROTECTION

Main provides shaded zone:
1. Disconnect to lockout
2. Overcurrent protection

Service Entrance
Main Design

No main leaves shaded area at greater risk to worker:
1. No disconnect to lockout
2. No premises overcurrent protection

Service Entrance
Six Disconnect Rule
No Main Design

Figure 11-25. These single-line diagrams illustrate systems with and without single main disconnect/overcurrent protection and indicate how each design affects the arc-flash hazard level.

to the right, which has no main overcurrent protection.

If a worker is performing a task within a feeder compartment with an exposed energized conductor, a main OCPD helps protect against arcing faults on the feeder device's line terminals and the equipment's main bus. An arc flash risk assessment associated with the main device must be performed because large ampere-rated OCPDs might permit high arc-flash incident energies. In most cases, the main OCPD can provide better protection than the utility OCPD, which is located on the transformer primary.

High-Speed Fuse Assembly

An engineered assembly installed at the secondary of a large transformer or at other locations that may potentially have high incident energy can greatly mitigate the arc-flash hazards. **See Figure 11-26.** These assemblies have been used in mining applications, where the transformer secondary has high load current requirements and high arc-flash hazards. They incorporate special-purpose high-speed fuses that are very current limiting. The resulting systems are engineered to carry the normal load currents, but respond very quickly to fault currents. They provide opportunities for electrical design teams on both

Figure 11-26. ENGINEERED ASSEMBLY

Figure 11-26. Engineered assemblies with high-speed fuses can be used to mitigate arc-flash hazards. Courtesy of PACE Engineers Group Pty Ltd.

greenfield and brownfield projects to minimize arc-flash hazards and can also lower a site's carbon footprint through the use of low-impedance transformers. Low-impedance transformers transformers provide greater available fault current, a factor that assists in the arc-flash mitigation operation provided by this assembly.

Sizing Underutilized Circuits with Lower-Ampere-Rated Fuses or Circuit Breakers

When the rated ampacity of a circuit is significantly larger than necessary, the actual load current under the maximum load conditions should be measured, and the most current-limiting fuses should be sized for the load. If an 800-ampere feeder to a motor control center draws only 320 amperes, for example, then 400-ampere current-limiting fuses could be considered. Downsizing fuses or circuit breakers is a strategy that can be utilized on some new systems or on retrofitted existing systems. **See Figure 11-27.**

Evaluating OCPDs in Existing Facilities for Interrupting Rating

Fuses, circuit breakers, and other OCPDs must be selected so that their interrupting rating is equal or greater than the available short-circuit current for the initial installation as well as over the life of the system.

Changes in an electrical system, including changes made to the supplying utility system, can result in higher available short-circuit currents, which may exceed the interrupting rating of existing OCPDs. When a service transformer is increased in kilovolt-ampere size or the transformer is replaced with a lower-impedance transformer, the available short-circuit current might increase to levels greater than the OCPDs' interrupting ratings.

In addition to *NEC* Section 110.9, which requires fuses and circuit breakers to have adequate interrupting ratings, *NFPA 70E* Section 210.5 and OSHA require that OCPDs have adequate interrupting ratings irrespective of the installation age of the system. §OSHA 29 CFR 1910.302 (b) requires some regulations to be followed irrespective of the original system installation date. One of those regulations is provided here.

> **§OSHA 29 CFR 1910.303(b)(4)**
> **Interrupting rating.** Equipment intended to interrupt current at fault levels shall have an interrupting rating sufficient for the nominal circuit voltage and the current that is available at the line terminals of the equipment. Equipment intended to interrupt current at other than fault levels shall have an interrupting rating at nominal circuit voltage sufficient for the current that must be interrupted.

Whenever system changes occur in a premise or changes are made by the utility that might increase the available short-circuit currents, the existing OCPDs must be reevaluated to determine whether they have sufficient interrupting rating. If a short-circuit analysis or an arc-flash risk assessment is performed for an existing facility, the employer must evaluate whether all the fuses and circuit breakers have sufficient interrupting rating for the available short-circuit current at their line terminals. Fuses or circuit breakers that do not have an adequate interrupting rating should be replaced with fuses or circuit breakers that have an adequate interrupting rating.

The definition for interrupting rating is essentially the maximum short-circuit current that a fuse or circuit breaker can interrupt at rated voltage. An OCPD that

Figure 11-27. DOWNSIZING

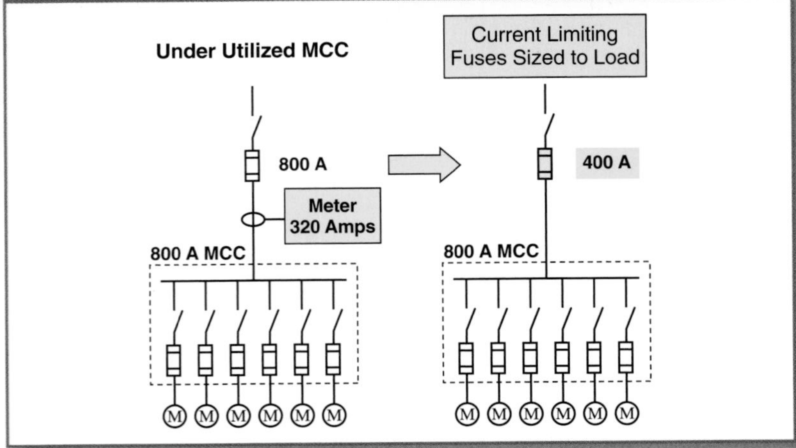

Under Utilized MCC

Current Limiting Fuses Sized to Load

800 A

Meter 320 Amps

800 A MCC

400 A

800 A MCC

Figure 11-27. On large-ampacity circuits that are lightly loaded, retrofit to lower ampacity OCPDs.

Figure 11-28. MISAPPLICATION OF FUSE INTERRUPTING RATING IS SAFETY HAZARD

For additional information, visit qr.njatcdb.org Item #1235.

Figure 11-28. A laboratory test illustrates what happens when Class H fuses, which have an interrupting rating of only 10,000 amperes, are subjected to a 50,000-ampere fault.

attempts to interrupt a short-circuit current beyond its interrupting rating can rupture violently. This misapplication condition can present an arc flash and arc-blast hazard, and the violent rupturing can initiate an arcing fault in other parts of the equipment. (Another chapter covers calculation of the maximum available short-circuit current.) Modern current-limiting fuses have interrupting ratings of 200,000 and 300,000 amperes,

which virtually eliminate this hazard contributor. However, renewable and Class H fuses have only a 10,000-ampere interrupting rating, some Class K fuses have a 50,000-ampere interrupting rating, and Class G fuses have a 100,000-ampere interrupting rating.

Circuit breakers have varying interrupting ratings, so they need to be assessed accordingly. **See Figures 11-28 and 11-29.** Although these tests depict a

Figure 11-29. MISAPPLICATION OF CIRCUIT BREAKER INTERRUPTING RATING IS SAFETY HAZARD

For additional information, visit qr.njatcdb.org Item #1236.

Figure 11-29. A laboratory test illustrates what happens when a circuit breaker with an interrupting rating of 14,000 amperes is subjected to a 50,000-ampere fault.

For additional
information, visit
qr.njatcdb.org
Item #1237.

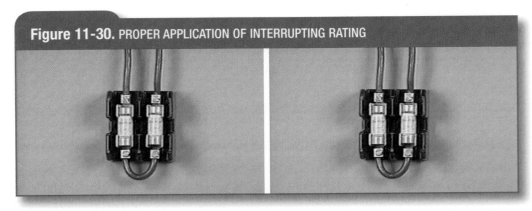

Figure 11-30. PROPER APPLICATION OF INTERRUPTING RATING

Figure 11-30. A laboratory test illustrates Class J, low-peak LPJ fuses safely interrupting a 50,000-ampere available short-circuit current. The LPJ fuses have an interrupting rating of 300,000 amperes. Left: Fuses are shown before a laboratory test. Right: Fuses are shown during and after a laboratory test to safely interrupt a short-circuit current.

violation of *NEC* Section 110.9, *NFPA 70E* 210.5, and OSHA §1910.303 (b)(4), they emphasize the importance of a proper interrupting rating for arc-flash protection and application of OCPDs. In a fraction of a second, sudden violence can occur.

A short-circuit current can be safety interrupted. **See Figure 11-30.** Note that in this laboratory test the fuses have an interrupting rating greater than the available short-circuit current; therefore, the current is safely interrupted.

"No Damage" Protection for Motor Controllers

Motor starters that are designated as Type 1 protection are susceptible to violent explosive damage from short-circuit currents. They can be the source of an arc flash or can produce ionized gas that initiates an arcing fault in an enclosure. If an employee needs to work within an enclosure that is not in an electrically safe work condition and that contains a motor starter protected by a Type 1 approach, he or she may be exposed to a serious safety hazard. Specifying Type 2 motor starter protection can reduce the risk, because the level of current limitation typically required to obtain Type 2 protection also provides for excellent arc-flash hazard reduction and minimizes the chance that the starter will be

the source of the arc flash or contribute to the initiation of an arcing fault.

Type 1 Protection. IEC 60947-4-1, "Type 1 Protection" (similar to requirements for listing motor starters to UL 508), requires that, under short-circuit conditions, the contactor or starter cause no danger to persons (with the enclosure door closed) or surroundings, but the contactor or starter might not be suitable for further service without repair and replacement of parts. Note that damage is allowed that may require partial or complete component replacement. It is possible for the overload devices to vaporize and the contacts to weld. Short-circuit protective devices interrupt the short-circuit current but are not required to prevent component damage. **See Figure 11-31.** These still photos were taken from a video of a motor starter tested to Type 1 criteria (with the door open so as to visualize the damage level). The heater elements vaporized and the contacts severely welded, contributing vaporized metal to the atmosphere. If a worker had any unprotected body parts near such an event, he or she might be injured.

Type 2 Protection. Using starters with OCPDs that provide Type 2 protection may mean that the starter is not a potential arc-flash contributor. UL 508E (Outline of Investigation) and IEC 60947-4-1, "Type 2 Protection," require that, under

Figure 11-31. TYPE 1 STARTER PROTECTION

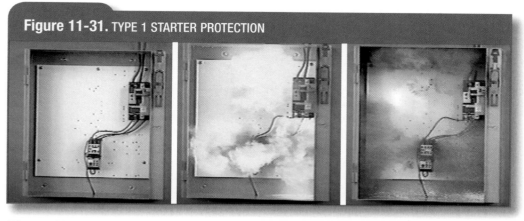

Figure 11-31. *The photos were taken before, during, and after testing a motor circuit protector (MCP) intended to provide motor branch-circuit protection for a 10-horsepower (hp) IEC starter with 22,000 amperes of short-circuit current available at 480 volts. The heater elements vaporized, and the contacts were severely welded. This could be a hazard if the door is open and a worker is near.*

short-circuit conditions, the contactor or starter cause no danger to persons (with the enclosure door closed) or the installation and be suitable for further use. No damage is allowed to either the contactor or the overload relay. Light contact welding is permitted, but contacts must be easily separable. "No damage" protection for the National Electrical Manufacturers Association (NEMA) and IEC motor start-ers can be provided only by a device that is able to limit the magnitude and the duration of short-circuit current. **See Figure 11-32.**

Fuses that typically meet the requirements for Type 2 "no damage" protection, as demonstrated by the results of the controller manufacturer's testing, include Class J, Class CF, Class CC, and Class RK1 fuses. As mentioned earlier in

Figure 11-32. TYPE 2 "NO DAMAGE" STARTER PROTECTION

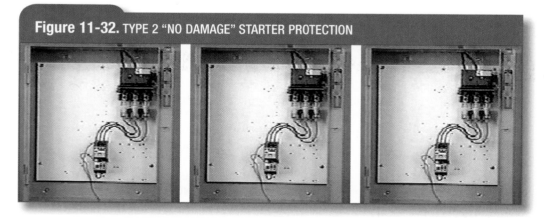

Figure 11-32. *These still photos were taken from a video of a motor starter tested to Type 2 criteria (with the door open so as to visualize the lack of a hazard). The photos were taken before, during, and after use of the same test circuit and same type of starter during short-circuit interruption as in Figure 11-24. The difference is that here Cooper Bussmann LPJ-SP Class J current-limiting fuses provide the motor branch-circuit protection. This level of protection reduces the risk for workers.*

this chapter, these fuses are very current limiting, which can protect the sensitive controller components.

Selective Coordination

Today, one of the most important parts of any installation is the electrical distribution system. Nothing can stop all activity, paralyze production, create inconvenience, disconcert people, and possibly cause a panic more than a power outage to critical loads.

Selective coordination is considered the act of isolating a faulted circuit from the remainder of the electrical system, thereby eliminating unnecessary power outages. The faulted circuit is isolated by the selective operation of only that OCPD closest to the fault condition. An adequately engineered system enables only the protective device nearest the fault to open, leaving the remainder of the system undisturbed and preserving continuity of service.

Personnel safety is enhanced in a selectively coordinated system because the Electrical Worker is not unnecessarily exposed to arc-flash hazards at upstream panels or switchboards. In a selectively coordinated system, the worker is exposed only at the level of circuit that incurred the problem. In a non-selectively coordinated system, the worker, while troubleshooting the open circuit, is required to work in upstream equipment, where arc-flash energies are often significantly higher. **See Figure 11-33.** In this single-line diagram, the arc-flash energy at a branch-circuit panel is 1.6 cal/cm². If an overcurrent condition on the branch circuit opens only the overcurrent device in the branch-circuit panel, the worker is exposed to 1.6 cal/cm² while troubleshooting the circuit. If the feeder overcurrent device unnecessarily opens due to cascading OCPDs, the worker is forced to work within the feeder panel and is unnecessarily exposed to a higher level of 6.7 cal/cm². If the main also opens unnecessarily, the worker is forced to work within the main panel, thereby exposing the worker to 12.3 cal/cm². Therefore, in this selectively coordinated system example, the worker is exposed to only 1.6 cal/cm², but in the nonselectively coordinated system, the worker is exposed to 12.3 cal/cm².

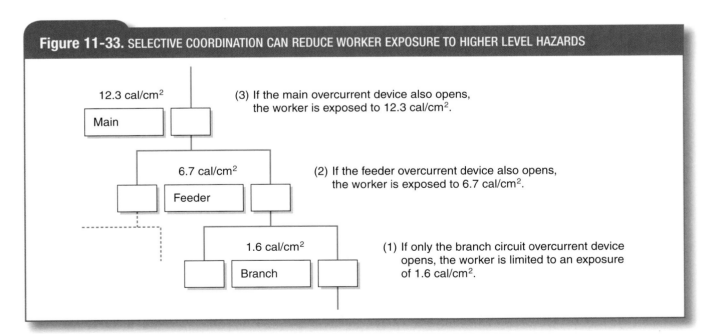

Figure 11-33. SELECTIVE COORDINATION CAN REDUCE WORKER EXPOSURE TO HIGHER LEVEL HAZARDS

12.3 cal/cm²

Main

(3) If the main overcurrent device also opens, the worker is exposed to 12.3 cal/cm².

6.7 cal/cm²

Feeder

(2) If the feeder overcurrent device also opens, the worker is exposed to 6.7 cal/cm².

1.6 cal/cm²

Branch

(1) If only the branch circuit overcurrent device opens, the worker is limited to an exposure of 1.6 cal/cm².

Figure 11-33. A lack of selective coordination can unnecessarily expose a worker to higher levels of arc-flash hazards.

ADDITIONAL DESIGN FOR SAFETY CONSIDERATIONS

Beyond overcurrent protective device safety design considerations, there are design techniques for new and existing systems that reduce or eliminate electrical hazards for workers. This section discusses a few of these options; however, a multitude of such design considerations exist.

Remote Monitoring

Specifying remote monitoring of voltage or current, or measurement of other vital electrical parameters, reduces exposure to electrical hazards by transferring the potentially hazardous troubleshooting activity from the actual live equipment to a display visible with the equipment doors closed, thereby enhancing safe work practices. **See Figure 11-34.** Such remote monitoring can be implemented in many ways. For example, the displays may be mounted so that they are readable from the equipment enclosure exterior; the equipment enclosure exterior may support "plug-in" diagnostic displays, instruments, or computers so that troubleshooting can be performed at the equipment with the doors closed; or remote computers may provide network access to the necessary data. These designs reduce the associated electrical hazards and reduce the number of times that required PPE must be worn by Electrical Workers.

Finger-Safe Products and Terminal Covers

One of the best ways to minimize exposure to electrical shock hazard is to use finger-safe products and nonconductive covers or barriers. ("Finger-safe" is a

70E Highlights

NFPA 70E *Annex O is helpful for generating designs that minimize exposure of workers to electrical hazards. The title of this annex is "Safety-Related Design Requirements."*

Figure 11-34. MONITORING DEVICES

Figure 11-34. *Remote monitoring devices for electrical system information (for example, voltage, ampacity) provide a means for workers to perform electrical diagnostics with the enclosure doors closed. Courtesy of Eaton Corporation.*

generic term but not a product standard criterion for evaluating electrical products' ability to minimize shock hazard.) Finger-safe products and covers reduce the chance of causing a shock or initiat-

ing an arcing fault. If all the electrical components are finger safe or covered, a worker has a much lower risk of making contact with an energized part (shock hazard). The risk of any conductive part

Figure 11-35. FINGER-SAFE DEVICES

CUBEFuses™

SAMI fuse covers

Fusible disconnects

Power distribution blocks

Class J fuse holders

Other fuse holders

Figure 11-35. Finger-safe devices can help minimize electrical hazards. Courtesy of Eaton's Bussmann Business.

falling across bare, energized conductive parts and creating an arcing fault (arc-flash hazard) is greatly reduced.

Several items have been developed that can help minimize shock hazards and minimize the initiation of an arcing fault. **See Figure 11-35.** All of these devices can reduce the chance that a worker, tool, or other conductive item will come in contact with a live part.

The International Protection (IP) Code rating is an IEC system, detailed in IEC 60529. **See Figure 11-36.** IP2X is often referred to as "finger safe," meaning that a probe the approximate size of a finger must not be able to access or make contact with hazardous energized parts. Principally, IEC 60529 defines the degree of protection provided by an enclosure (barriers/guards) classified under the IP Code and the testing conditions required to meet these classifications. IP20-rated products do not offer any protection against liquids. UL product standards are lagging in the adoption of evaluating enclosures and components for shock hazards; however, some UL product standards are starting to evaluate electrical devices (components) for compliance with IP20. In some cases, the manufacturers of components may make self-certified IP20 claims.

Workers should understand the concept underlying the IP20 rating and its benefits and limitations. Some component products with IP20 claims have a

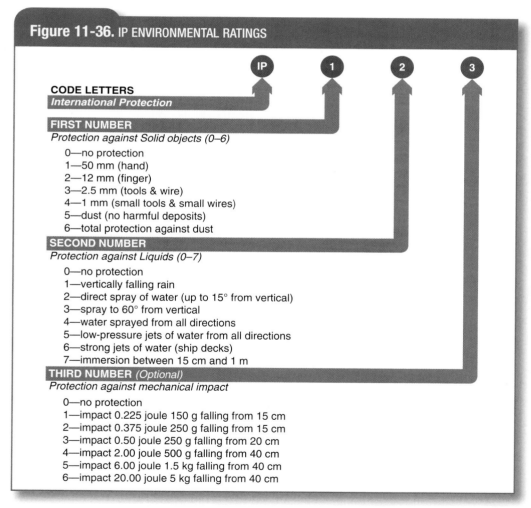

Figure 11-36. IP ENVIRONMENTAL RATINGS

CODE LETTERS
International Protection

FIRST NUMBER
Protection against Solid objects (0–6)

  0—no protection
  1—50 mm (hand)
  2—12 mm (finger)
  3—2.5 mm (tools & wire)
  4—1 mm (small tools & small wires)
  5—dust (no harmful deposits)
  6—total protection against dust

SECOND NUMBER
Protection against Liquids (0–7)

  0—no protection
  1—vertically falling rain
  2—direct spray of water (up to 15° from vertical)
  3—spray to 60° from vertical
  4—water sprayed from all directions
  5—low-pressure jets of water from all directions
  6—strong jets of water (ship decks)
  7—immersion between 15 cm and 1 m

THIRD NUMBER *(Optional)*
Protection against mechanical impact

  0—no protection
  1—impact 0.225 joule 150 g falling from 15 cm
  2—impact 0.375 joule 250 g falling from 15 cm
  3—impact 0.50 joule 250 g falling from 20 cm
  4—impact 2.00 joule 500 g falling from 40 cm
  5—impact 6.00 joule 1.5 kg falling from 40 cm
  6—impact 20.00 joule 5 kg falling from 40 cm

Figure 11-36. IP environmental ratings for enclosures (IEC 60529) can be used to minimize electrical hazards.

"conditional" IP20 rating. For example, the conductor termination may be considered IP20 rated if the conductor is prepared and installed in the terminal properly. This same terminal may be IP20 rated for the larger AWG conductors for which the terminal is rated, but not IP20 rated for the smaller AWG conductors for which the terminal is rated (a bare finger could contact bare energized metal).

Isolating the Circuit: Installation of an "In-Sight" Disconnect for Each Motor

Electrical systems should be designed to support maintenance, with easy, safe access to the equipment. Their design should provide for isolating equipment for repair purposes, with a disconnecting means for implementing lockout/tagout procedures. A sound design provides disconnecting means at all motor loads, in addition to disconnecting means required within sight of the controller location that can be locked in the open position. Disconnecting means at the motor provide improved isolation and safety for maintenance as well as an emergency disconnect.

Horsepower-rated disconnects should be installed within sight (visible and within 50 feet) of every motor or driven machine. **See Figure 11-37.** The provision for locking or adding a lock to the disconnecting means shall be installed on or at the switch or circuit breaker used as the disconnecting means and shall remain in place with or without the lock installed. The *NEC* uses the following terms for the same purpose: "in-sight from," "within sight from", or "within sight of" to mean that the specified equipment is to be visible and not more than 15 m (50 ft) distant from the other.

An in-sight motor disconnect is more likely to be used by a worker for the lockout procedure to put equipment in an electrically safe work condition prior to doing work on the equipment. Such a disconnect generally is required even if the disconnect within sight of the controller can be locked out. Some exceptions exist for specific industrial applications.

Breaking up Large Circuits into Smaller Circuits

In the design phase, if preliminary arc-flash risk assessment indicates that equipment in high-ampacity circuits is associated with high arc-flash hazards, it may be feasible to divide the loads and supply the various pieces of equipment by multiple smaller-ampacity feeders. In many cases, very large ampere-rated fuses and circuit breakers let through too much energy for a practical PPE arc rating.

A 1,600-ampere circuit, for example, might potentially be broken into two 800-ampere circuits. An analysis of the arc-flash incident energy available in each circuit would then generally indicate that two 800-ampere circuits would be better than one 1,600-ampere circuit.

For specific situations, an arc-flash risk assessment should be completed, as variables can affect the outcome. This is especially beneficial when using current-limiting protective devices, because the lower ampere-rated devices are typically more current-limiting and, therefore, can better reduce the arc-flash risk assessment. **See Figure 11-38.**

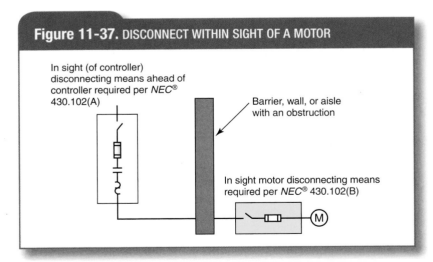

Figure 11-37. DISCONNECT WITHIN SIGHT OF A MOTOR

In sight (of controller) disconnecting means ahead of controller required per *NEC*® 430.102(A)

Barrier, wall, or aisle with an obstruction

In sight motor disconnecting means required per *NEC*® 430.102(B)

Figure 11-37. A disconnect should be installed within sight of a motor.

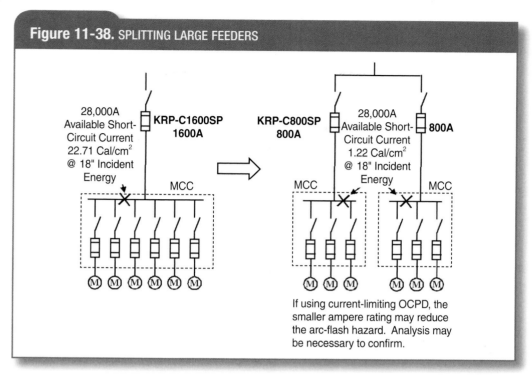

Figure 11-38. SPLITTING LARGE FEEDERS

28,000A Available Short-Circuit Current 22.71 Cal/cm^2 @ 18" Incident Energy

KRP-C1600SP 1600A

MCC

KRP-C800SP 800A

28,000A Available Short-Circuit Current 1.22 Cal/cm^2 @ 18" Incident Energy

800A

MCC

MCC

If using current-limiting OCPD, the smaller ampere rating may reduce the arc-flash hazard. Analysis may be necessary to confirm.

Figure 11-38. Lower ampere-rated devices are typically more current limiting and, therefore, can better reduce the arc-flash hazard.

Remote Opening and Closing

Opening and closing large switches and circuit breakers have caused serious arc-flash incidents when these devices failed while being operated. By opening and closing large switches and circuit breakers remotely, a worker might control the operation from a safe distance. If an arc-flash incident should occur in this scenario, the worker will not be exposed to the hazard. **See Figure 11-39.**

Arc Flash Preventative MCCs

See Figures 11-40 and 11-41. Some motor Control Centers (MCC) are available with many design features which improve the safety for when maintenance is required. With this motor control center (MCC) the buckets can be racked in and out remotely.

What makes for arc flash preventative MCCs? Key strategies are used to help safeguard employees against injuries from

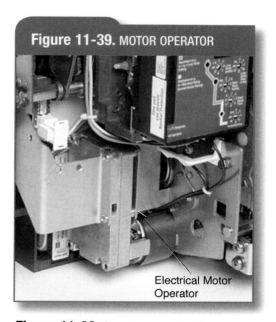

Figure 11-39. MOTOR OPERATOR

Electrical Motor Operator

Figure 11-39. A motor operator for remote opening and closing of a circuit breaker can permit a worker to operate a circuit breaker from a long distance. Courtesy of Eaton Corporation.

Figure 11-40. MOTOR CONTROL CENTER

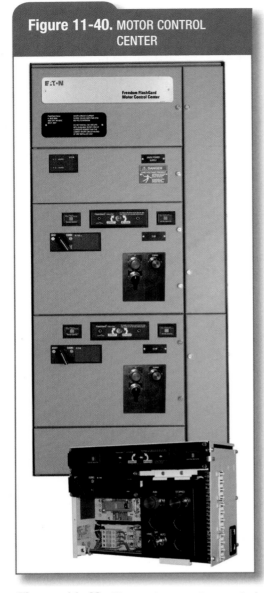

Figure 11-40. *Shown is a motor control center (MCC) which has the feature to rack the MCC buckets in and out with a remote racking tool shown in Figure 11-41.* Courtesy of Eaton Corporation.

Figure 11-41. REMOTE TRACKING DEVICE

Figure 11-41. *This motorized remote racking device is a companion to the MCC in Figure 11-40; this device permits racking the MCC buckets in and out remotely.* Courtesy of Eaton Corporation.

electric shock, arc flash burns and arc blasts:

1. Multiple insulation and isolation features enable arc flash prevention.
2. Unlike conventional MCCs, arc-preventative MCC design enables units to be disconnected and re-connected to the vertical bus with the door closed; maintaining a closed door during these operations increases operator safety. Additionally, remote racking device permits the maintenance personnel to stand at a safe distance.
3. A series of safety interlocks ensures that doors cannot be opened and that units cannot be removed from the structure while the stabs are connected to the vertical bus.
4. Each unit contains visual indicators that report the position of the isolation shutters and the stabs, providing maintenance personnel with additional assurance that dangerous voltages are not present inside the unit when service is required.

Arc-Resistant (Arc-Diverting) Medium-Voltage Switchgear

Arc-resistant switchgear can be installed to withstand internal arcing faults. Arc-resistant equipment typically is designed with stronger door hinges and latches, better door gaskets, and hinged enclosure top venting panels. The underlying concept focuses on diversion of the resultant explosive hot gases and pressure from an internal arcing fault via the hinged enclosure top panels. If the switchgear is installed indoors, then a means of exhausting the hot gases to the outside of the building, such as ducts, is required.

Arc-resistant equipment is rated to withstand specific levels of internal arcing faults with all the doors closed and latched. The rating does not apply with any door opened or any cover removed. Therefore, arc-resistant equipment does not protect a worker who is performing a task with an open door or panel.

The term "arc-resistant" is a bit misleading. The internal switchgear must withstand an internal arcing fault and, therefore, the sheet metal and other components of the equipment must resist or withstand a specified arcing fault. However, a major feature of this equipment is diversion of the arcing fault by-products (that is, hot ionized gases and blast) via the enclosure top panels. This feature helps to prevent the arcing fault from blowing open the doors or side panels and venting the arcing fault by-products where a worker might be standing. **See Figure 11-42.**

Service Panel with Isolated Barrier

See Figure 11-43. For these service entrance panels, the area in the panel where service conductors terminate to the panel main disconnect or main lugs is isolated by a sheet metal barrier. This feature enhances safety by preventing inadvertent contact with the conductors and terminations in this isolated area. This barrier is not intended for arc flash protection, however, it does reduce the probability that an arc flash incident will occur.

Resistance-Grounded Systems

Electrical systems can be designed with resistance-grounded wye systems to increase system reliability and reduce the

Figure 11-42. ARC-RESISTANCE SWITCHGEAR WITH VENTING DUCTS ABOVE GEAR

Figure 11-42. Arc-resistant switchgear diverts the arcing fault gases and pressure out the top of the gear if all doors are closed and latched properly. Courtesy of Eaton Corporation.

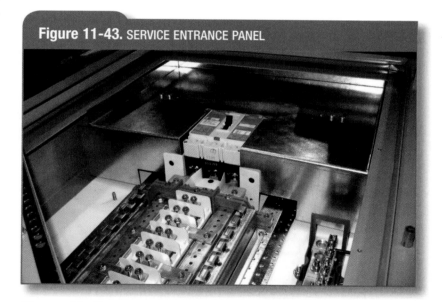

Figure 11-43. SERVICE ENTRANCE PANEL

Figure 11-43. Some panels suitable for service entrance applications have the added feature that the area of the panel where the service conductors terminate to the main OCPD/disconnect is isolated by the sheet metal barrier. Courtesy of Eaton Corporation.

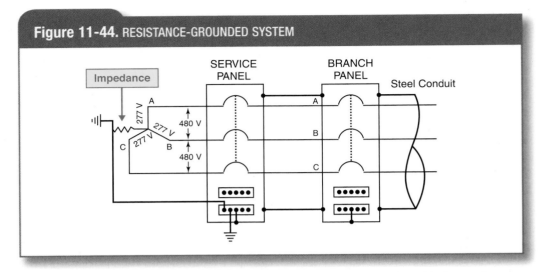

Figure 11-44. *In a resistance-grounded system, the resistor is positioned between the center tap of the wye transformer and the ground.*

probability of arc-flash incidents. **See Figure 11-44.** In this type of installation, a resistor is intentionally inserted between the center point of the wye of the transformer and the ground, so that the ground short-circuit current is limited by the inserted resistor. For three-phase systems, this type of system can only be used for a 3-wire system; that is, it is not permitted by the NEC for systems such as 277V/480V, 3-phase, 4-wire.

This type of system can reduce the probability that dangerous or destructive

arcing faults will occur. For example, with a resistance-grounded wye system, if a worker's screwdriver slips, simultaneously touching an energized bare-phase termination and the enclosure, a high-energy arc fault would not be initiated. **See Figure 11-45.** Instead, the fault current would typically be limited to only a few amperes by the resistor intentionally inserted in the ground return path. This limited fault current is insufficient to create the vaporized metal that would initiate or sustain a 3-phase arcing fault.

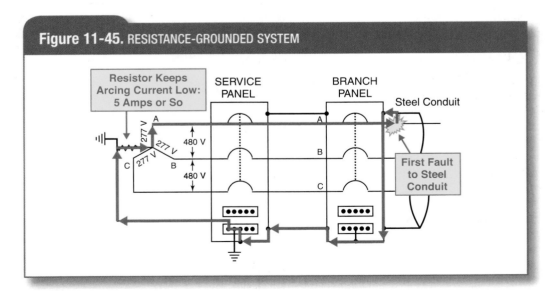

Figure 11-45. *Because of the resistance, a fault occurring from phase to ground does not cause the OCPD to open, since the resistor limits the fault current to approximately 5 A.*

Figure 11-46. RESISTANCE-GROUNDED SYSTEM

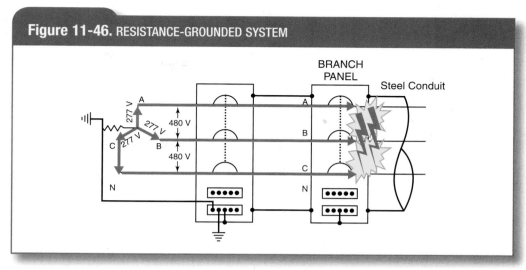

Figure 11-46. *A high energy release arcing fault incident is still possible. If a fault occurs between only two phases or all three phases, the resistor is not in the fault current path and high fault currents are possible. So a arc flash risk assessment is required for this possibility.*

See **Figure 11-46.** A high resistance grounded system does not eliminate the arc-flash hazard, however. If the worker's screwdriver simultaneously touches the energized bare terminations of two phases, a high-level arcing fault might occur because the resistor then is not involved in the faulted current path. If so, it can quickly escalate into a 3-phase arcing fault.

Cautionary Note: **See Figure 11-47.** If a designer is planning high-resistance/grounded wye systems or retrofitting an existing solidly grounded wye system, he or she must consider the single-pole

Figure 11-47. RESISTANCE-GROUNDED SYSTEM

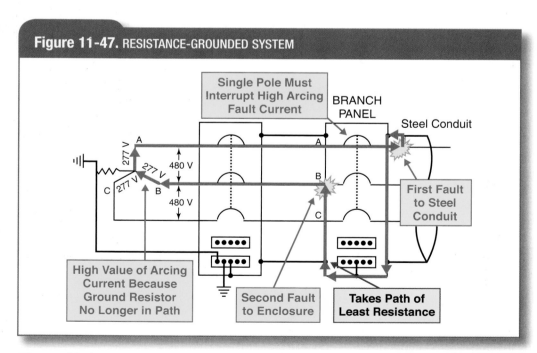

Figure 11-47. *The first fault to ground must be removed by Electrical Workers before a second phase goes to ground, or a significant short-circuit current can occur across one pole of the branch-circuit device.*

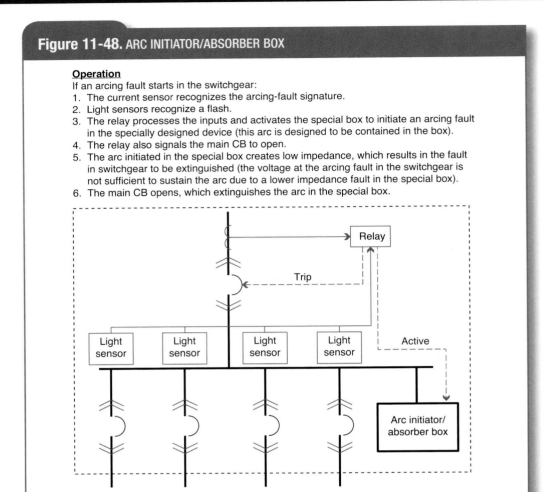

Figure 11-48. ARC INITIATOR/ABSORBER BOX

Operation
If an arcing fault starts in the switchgear:
1. The current sensor recognizes the arcing-fault signature.
2. Light sensors recognize a flash.
3. The relay processes the inputs and activates the special box to initiate an arcing fault in the specially designed device (this arc is designed to be contained in the box).
4. The relay also signals the main CB to open.
5. The arc initiated in the special box creates low impedance, which results in the fault in switchgear to be extinguished (the voltage at the arcing fault in the switchgear is not sufficient to sustain the arc due to a lower impedance fault in the special box).
6. The main CB opens, which extinguishes the arc in the special box.

Figure 11-48. *An arc initiator/absorber box can reduce the arc-flash hazard.*

interrupting capabilities of any circuit breakers and self-protected starters to be installed or already installed. Slash voltage-rated circuit breakers and self-protected starters cannot be used on any system other than solidly grounded wye systems.

Arc Initiator/Absorber Box

An arc initiator/absorber box is a specially designed box made to safely contain an arcing fault when it is used with a relay, light sensors, current sensors, and a shunt-trip main circuit breaker. With this approach, if an arcing fault starts in the switchgear, it is detected, and the special box initiates an arc in the box (crow bar), which then extinguishes the arcing fault in the switchgear. **See Figure 11-48.**

Summary

Electrical system design, whether for a new system or an upgrade, can have a significant impact on personnel safety. The choices made during the design stage play an important role in reducing or eliminating exposure to electrical hazards. Designers must carefully select equipment and circuit designs that provide maximum protection for workers. The selection of the type and characteristics of OCPDs has a significant effect on the arc-flash hazard; in addition, other design techniques can affect the arc-flash hazard and other hazards. Many design techniques other than those presented in this chapter are possible.

Review Questions

1. By __?__, there is the opportunity to eliminate or minimize work exposure to electrical hazards before a worker ever has a chance to become exposed to them.
   a. designing for safety
   b. performing arc flash hazard analysis
   c. using insulated tools
   d. using PPE

2. __?__ are the "safety valves" that may limit the magnitude of the fault current that flows and reduce the time duration of the current, thereby mitigating the arc-flash hazard to some extent.
   a. Finger-safe products
   b. Overcurrent protective devices
   c. PPE
   d. Transformers

3. Circuit breakers and fuses that can significantly reduce the magnitude of fault current and clear in ½ cycle or less for some range of overcurrents are listed as __?__.
   a. current-limiting
   b. electrical equipment
   c. non-current limiting
   d. overcurrent protective devices

4. The longer an OCPD takes to clear a given arcing fault current, the __?__ the arc-flash hazard.
   a. greater
   b. less the incident energy of
   c. lesser
   d. safer

5. __?__ do not require maintenance other than visual examination and protection from damage due to external conditions.
   a. Circuit breakers
   b. Current-limiting fuses
   c. Protective relays
   d. Vacuum interrupters

6. __?__ for circuit breakers, where short-time delay is used (and no instantaneous trip setting), reduces incident energy exposure levels by turning off the short-time delay function and invoking an instantaneous trip while a person is working within the arc-flash protection boundary.
   a. Arc flash-reducing maintenance switching
   b. Electronic sensing
   c. Lubrication
   d. Remote racking

7. Fuses and circuit breakers must be selected so that their interrupting rating is equal to or __?__ the available short-circuit current for the initial installation as well as over the life of the system.
   a. greater than
   b. less than
   c. the same as
   d. weaker than

8. __?__ is a design method that is an effective arc-flash mitigation means for equipment protected by OCPDs that are too large in ampacity to quickly respond to the arcing current.
   a. A ground fault relay
   b. An center taped transformer
   c. A potential transformer
   d. Arc reduction maintenance system

9. __?__ provide a means for workers to perform electrical diagnostics with the enclosure doors closed.
   a. Arc-reduction maintenance systems
   b. Finger-safe products
   c. Insulated tools
   d. Remote monitoring devices

10. Disconnecting means at the motor provide improved __?__ and __?__ for maintenance as well as an emergency disconnect.
    a. fault protection/visibility
    b. isolation/safety
    c. voltage regulation/safety
    d. voltage regulation/visibility

Appendix

Figure A-1. 3-PHASE TRANSFORMER FULL LOAD CURRENTS

3-Phase Voltage (Line-to-Line)	3-Phase Transformer kVA Rating								
	150	**167**	**225**	**300**	**500**	**750**	**1000**	**1500**	**2000**
208	417	464	625	833	1388	2080	2776	4164	5552
220	394	439	592	788	1315	1970	2630	3940	5260
240	362	402	542	722	1203	1804	2406	3609	4812
440	197	219	296	394	657	985	1315	1970	2630
460	189	209	284	378	630	945	1260	1890	2520
480	181	201	271	361	601	902	1203	1804	2406
600	144	161	216	289	481	722	962	1444	1924

Figure A-1. 3-Phase Transformer: Full Load Current Rating (in amperes) values can be used in lieu of the equation shown in Step 1 of the 3-phase short-circuit current calculation procedure.

Figure A-2. "C" VALUES FOR COPPER CONDUCTORS

AWG or kcmil	Copper Conductors											
	Three Single Conductors Conduit						Three-Conductor Cable Conduit					
	Steel			Nonmagnetic			Steel			Nonmagnetic		
	600 V	5 kV	15 kV	600 V	5 kV	15 kV	600 V	5 kV	15 kV	600 V	5 kV	15 kV
14	389	—	—	389	—	—	389	—	—	389	—	—
12	617	—	—	617	—	—	617	—	—	617	—	—
10	981	—	—	982	—	—	982	—	—	982	—	—
8	1557	1551	—	1559	1555	—	1559	1557	—	1560	1558	—
6	2425	2406	2389	2430	2418	2407	2431	2425	2415	2433	2428	2421
4	3806	3751	3696	3826	3789	3753	3830	3812	3779	3838	3823	3798
3	4774	4674	4577	4811	4745	4679	4820	4785	4726	4833	4803	4762
2	5907	5736	5574	6044	5926	5809	5989	5930	5828	6087	6023	5958
1	7293	7029	6759	7493	7307	7109	7454	7365	7189	7579	7507	7364
1/0	8925	8544	7973	9317	9034	8590	9210	9086	8708	9473	9373	9053
2/0	10,755	10,062	9390	11,424	10,878	10,319	11.245	11,045	10,500	11,703	11,529	11,053
3/0	12,844	11,804	11,022	13,923	13,048	12,360	13,656	13,333	12,613	14,410	14,119	13,462
4/0	15,082	13,606	12,543	16,673	15,351	14,347	16,392	15,890	14,813	17,483	17,020	16,013
250	16,483	14,925	13,644	18,594	17,121	15,866	18,311	17,851	16,466	19,779	19,352	18,001
300	18,177	16,293	14,769	20,868	18,975	17,409	20,617	20,052	18,319	22,525	21,938	20,163
350	19,704	17,385	15,678	22,737	20,526	18,672	22,646	21,914	19,821	24,904	24,126	21,982
400	20,566	18,235	16,366	24,297	21,786	19,731	24,253	23,372	21,042	26,916	26,044	23,518
500	22,185	19,172	17,492	26,706	23,277	21,330	26,980	25,449	23,126	30,096	28,712	25,916
600	22,965	20,567	17,962	28,033	25,204	22,097	28,752	27,975	24,897	32,154	31,258	27,766
750	24,137	21,387	18,889	29,735	26,453	23,408	31,051	30,024	26,933	34,605	33,315	29,735
1000	25,278	22,539	19,923	31,491	28,083	24,887	33,864	32,689	29,320	37,917	35,749	31,959

See Note at bottom of Figure 8-17.

Figure A-2. "C" values for copper conductors are used in the Step 4 equation of the 3-phase short-circuit current calculation procedure.

Figure A-3. "C" VALUES FOR ALUMINUM CONDUCTORS

AWG or kcmil	Aluminum Conductors											
	Three Single Conductors Conduit						Three-Conductor Cable Conduit					
	Steel			Nonmagnetic			Steel			Nonmagnetic		
	600 V	5 kV	15 kV	600 V	5 kV	15 kV	600 V	5 kV	15 kV	600 V	5 kV	15 kV
14	237	—	—	237	—	—	237	—	—	237		
12	376	—	—	376	—	—	376	—	—	376		
10	599	—	—	599	—	—	599	—	—	599		
8	951	950	—	952	951	—	952	951	—	952	952	
6	1481	1476	1472	1482	1479	1476	1482	1480	1478	1482	1481	1479
4	2346	2333	2319	2350	2342	2333	2351	2347	2339	2353	2350	2344
3	2952	2928	2904	2961	2945	2929	2963	2955	2941	2966	2959	2949
2	3713	3670	3626	3730	3702	3673	3734	3719	3693	3740	3725	3709
1	4645	4575	4498	4678	4632	4580	4686	4664	4618	4699	4682	4646
1/0	5777	5670	5493	5838	5766	5646	5852	5820	5717	5876	5852	5771
2/0	7187	6968	6733	7301	7153	6986	7327	7271	7109	7373	7329	7202
3/0	8826	8467	8163	9110	8851	8627	9077	8981	8751	9243	9164	8977
4/0	10,741	10,167	9700	11,174	10,749	10,387	11,185	11,022	10,642	11,409	11,277	10,969
250	12,122	11,460	10,849	12,862	12,343	11,847	12,797	12,636	12,115	13,236	13,106	12,661
300	13,910	13,009	12,193	14,923	14,183	13,492	14,917	14,698	13,973	15,495	15,300	14,659
350	15,484	14,280	13,288	16,813	15,858	14,955	16,795	16,490	15,541	17,635	17,352	16,501
400	16,671	15,355	14,188	18,506	17,321	16,234	18,462	18,064	16,921	19,588	19,244	18,154
500	18,756	16,828	15,657	21,391	19,503	18,315	21,395	20,607	19,314	23,018	22,381	20,978
600	20,093	18,428	16,484	23,451	21,718	19,635	23,633	23,196	21,349	25,708	25,244	23,295
750	21,766	19,685	17,686	25,976	23,702	21,437	26,432	25,790	23,750	29,036	28,262	25,976
1000	23,478	21,235	19,006	28,779	26,109	23,482	29,865	29,049	26,608	32,938	31,920	29,135

Note: The values for Figures 8-16 and 8-17 are equal to 1 over the impedance per 1000 feet and based upon resistance and reactance values found in IEEE Std 241-1990 (Gray Book), IEEE Recommended Practice for Electric Power Systems in Commercial Buildings, and IEEE Std 242-1986 (Buff Book), IEEE Recommended Practice for Protection and Coordination of Industrial and Commercial Power Systems. Where resistance and reactance values differ or are not available, the Buff Book values have been used. The values for reactance in determining the "C" value at 5 and 15 kV are from the Gray Book only (values for 14-10 AWG at 5 kV and 14-8 AWG at 15 kV are not available, and values for 3 AWG have been approximated).

Figure A-3. "C" values for aluminum conductors are used in the Step 4 equation of the 3-phase short-circuit current calculation procedure.

Figure A-4. "C" VALUES FOR BUSWAY

Ampacity	Busway					
	Plug-In		Feeder		High Impedance	
	Copper	Aluminum	Copper	Aluminum	Copper	
225	28,700	23,000	18,700	12,000	—	
400	38,900	34,700	23,900	21,300	—	
600	41,000	38,300	36,500	31,300	—	
800	46,100	57,500	49,300	44,100	—	
1000	69,400	89,300	62,900	56,200	15,600	
1200	94,300	97,100	76,900	69,900	16,100	
1350	119,000	104,200	90,100	84,000	17,500	
1600	129,900	120,500	101,000	90,900	19,200	
2000	142,900	135,100	134,200	125,000	20,400	
2500	143,800	156,300	180,500	166,700	21,700	
3000	144,900	175,400	204,100	188,700	23,800	
4000	—	—	277,800	256,400	—	

Note: These values are equal to 1 over the impedance per foot for impedance in a survey of industry.

Figure A-4. "C" values for busway are used in the Step 4 equation of the 3-phase short-circuit current calculation procedure.

Figure A-5. MULTIPLIER = 1/(1 + F)

f	M	f	M
0.01	0.99	1.5	0.4
0.02	0.98	1.75	0.36
0.03	0.97	2	0.33
0.04	0.96	2.5	0.29
0.05	0.95	3	0.25
0.06	0.94	3.5	0.22
0.07	0.93	4	0.2
0.08	0.93	5	0.17
0.09	0.92	6	0.14
0.1	0.91	7	0.13
0.15	0.87	8	0.11
0.2	0.83	9	0.1
0.25	0.8	10	0.09
0.3	0.77	15	0.06
0.35	0.74	20	0.05
0.4	0.71	30	0.03
0.5	0.67	40	0.02
0.6	0.63	50	0.02
0.7	0.59	60	0.02
0.8	0.55	70	0.01
0.9	0.53	80	0.01
1.0	0.5	90	0.01
1.2	0.45	100	0.01

* $M = 1/(1 + f)$

Figure A-5. This table can be used to determine "M" (multiplier)* rather than doing the calculation in Step 5 or Step B when "f" is known from Step 4 or Step A.

Figure A-6. HIGHEST AVAILABLE SHORT-CIRCUIT CURRENT FOR TRANSFORMERS FROM SURVEY

Voltage and Phase	kVA	Full Load Amps	% Impedance [tt] (Nameplate)	Short-Circuit Amps [t]	Voltage and Phase	kVA	Full Load Amps	% Impedance [tt] (Nameplate)	Short-Circuit Amps [t]
277/480 3-phase**	112.5	135	1	15,000	120/208 3-phase**	25	69	1.6	4,791
	150	181	1.2	16,759		50	139	1.6	9652
	225	271	1.2	25,082		75	208	1.11	20,821
	300	361	1.2	33,426		112.5	278	1.11	27,828
	500	601	1.3	51,362		150	416	1.07	43,198
	750	902	3.5	28,410		225	625	1.12	62,004
	1000	1203	3.5	38,180		300	833	1.11	83,383
	1500	1804	3.5	57,261		500	1388	1.24	124,373
	2000	2406	5	53,461		750	2082	3.5	66,095
	2500	3007	5	66,822		1000	2776	3.5	88,167
	–	–	–	–		1500	4164	3.5	132,190
	–	–	–	–		2000	5552	5	123,377
	–	–	–	–		2500	6950	5	154,444

**3-phase short-circuit currents based on "infinite" primary.
[tt]UL-listed transformers 25 kVA or greater have a ±10% impedance tolerance. Short-circuit amps reflect a "worst-case" condition.
[t]Fluctuations in system voltage will affect the available short-circuit current. For example, a 10% increase in system voltage will result in a 10% increase in the available short-circuit currents shown in the table.

Figure A-6. *These generalized worst-case or highest available short-circuit current for 3-phase transformers of given kVA and secondary voltage provide some guidance. The lowest impedance of transformers is assumed based on a survey, infinite primary short-circuit current available, as well as worst-case adjustment values for transformer impedance tolerance and system voltage fluctuations.*

Figure A-7. SHORT-CIRCUIT CURRENTS AS A PERCENTAGE OF 3-PHASE BOLTED SHORT-CIRCUIT CURRENTS

3-phase (L-L-L) bolted fault	100%
Line-to-line (L-L) bolted fault	87%
Line-to-ground (L-G) bolted fault	25%–125%* (use 100% at transformer, 50% elsewhere)
Line-to-neutral (L-N) bolted fault	25%–125%* (use 100% at transformer, 50% elsewhere)
Three-phase (L-L-L) arcing fault (480 V)	89%** (maximum)
Line-to-line (L-L) arcing fault (480 V)	74%** (maximum)
Line-to-ground (L-G) arcing fault (480 V)	38%** (minimum)

*Typically line-ground bolted fault current much lower than the bolted 3-phase fault current, but the bolted line-ground fault current can exceed the 3-phase bolted fault current if it is near the transformer terminals. Bolted line-ground fault current will normally be between 25% to 125% of the 3-phase bolted fault value.
**These arcing fault values as a percentage of the 3-phase bolted short-circuit currents are rough estimates. See IEEE 1584, Guide for Performing Arc-Flash Hazard Calculations, for equations to calculate the arcing current when performing an arc flash hazard analysis.
Data source: IEEE Standard 241-1990, Recommended Practice for Electric Power Systems in Commercial Buildings, Table 63.

Figure A-7. *To provide estimates, various types of short-circuit currents as a percentage of 3-phase bolted fault current are shown.*

Index

NOTES

NOTES

NOTES